MEXICAN OIL AND NATURAL GAS

MEXICAN OIL AND NATURAL GAS

Political, Strategic, and Economic Implications

Richard B. Mancke

PRAEGER PUBLISHERS
Praeger Special Studies

New York • London • Sydney • Toronto

Library of Congress Cataloging in Publication Data

Mancke, Richard B   1943-
   Mexican oil and natural gas.

   Bibliography: p.
   1. Petroleum industry and trade--Mexico.
2. Gas industry--Mexico. 3. Gas, Natural--
Mexico. I. Title.
HD9574.M6M32   338.2'7'280972   78-27095
ISBN 0-03-048451-0

PRAEGER PUBLISHERS, PRAEGER SPECIAL STUDIES
383 Madison Avenue, New York, N.Y., 10017, U.S.A.

Published in the United States of America in 1979
by Praeger Publishers,
A Division of Holt, Rinehart and Winston, CBS Inc.

0 038 98765432

ACKNOWLEDGMENTS

Suspecting that it might be a useful pedagogical device to have a select group of graduate students in international relations help to produce a coordinated, publishable study of an important energy policy issue, I decided to have the Fletcher School of Law and Diplomacy's spring 1978 seminar on International Energy and Environmental Problems study the political, strategic, and economic ramifications of the huge petroleum discoveries in southeastern Mexico. This book is an outgrowth of that seminar. The participants in the seminar were Arthur Boley, Stephen Davis, Leslie Fenlon, Robert Fisher, Dean Goodermote, Jamie Kirkpatrick, Michael Metz, Peter Milone, Bruce Mohl, Ernesto Molina, Jr., Ann Moser, Mark Mueller, Catherine Rau, Thomas Sadler, Erik Sivesind, and Jesús Velasco Siles. Their specific contributions are described in the chapter footnotes.

In addition to the seminar's participants, several others deserve special thanks. Jesús Augustin Velasco Siles and Dr. Manuel Velasco Suarez (former governor of Chiapas) made arrangements for my visit to the Reforma area and to Mexico City. While in Mexico, the comments that I elicited during presentations before groups at El Colegio de Mexico, the Harvard Club of Mexico, and President Portillo's Group of the Advisers proved especially valuable. I am also indebted to Professor Neil Jacoby, Dr. Armand Hammer, and to my colleagues Professors William Barnes and Hewson Ryan (U.S. ambassador, retired) for discussing their assessments of the problems and potentials of Mexican oil and providing me with a steady stream of related information. Dean Goodermote prepared the maps and provided general assistance. Barbara Fennessy typed the manuscript. Finally, special thanks are owed to Barbara Hobbie for reading the entire manuscript and providing valuable editorial and substantive criticism.

CONTENTS

LIST OF TABLES AND FIGURE

LIST OF MAPS

LIST OF ACRONYMS

AIOC — American Independent Oil Company
API — American Petroleum Institute
CNC — Confederación Nacional Campesina
CNOP — Confederación Nacional de Organizaciones Populares
IMF — International Money Fund
LNG — Liquified Natural Gas
MBD — Million Barrels Per Day
Mcf — One Thousand Cubic Feet
PRI — Partido Revolucionario Institucional
OAPEC — Organization of Arab Petroleum Exporting Countries
PEMEX — Petróleos Mexicanos (Mexican National Oil Company)
STPRM — Sindicato de Trabajadores Petróleros de la República
(Union of the Oil Workers of the Republic)

1

THE PROMISE AND PROBLEMS
OF MEXICAN OIL:
AN OVERVIEW

The swamps and floodlands of the southeastern Mexican states of Chiapas and Tabasco had long been regarded by most geologists as a petroleum backwater. Consequently, oil industry observers were surprised when Mexico's national oil company, Petróleos Mexicanos (or Pemex), struck deposits of commercially recoverable oil while drilling the region's first deep wildcats in 1972. In the early years following Pemex's discovery, world oil observers attached relatively little importance to what was to be labeled the *Reforma Trend*. Thus, in February 1975, the *Wall Street Journal* expressed modest optimism when it reported that "astute observers in and out of Mexico wouldn't be surprised if oil reserves in the Reforma region eventually prove to total at least two billion and possibly four billion barrels or more."[1] Three months later, the authorative oil industry trade journal, *The Oil and Gas Journal*, ran a less optimistic headline: "Mexico's Crude Exporting Role May Be Short-Lived."[2] At 4 billion barrels, the Reforma Trend would have qualified as a large discovery—roughly half as large as the Prudhoe Bay field on Alaska's North Slope. However, it would not have been large enough to trigger substantial changes in the course of the world oil business. This assessment started to change rapidly in late 1976, as Pemex (apparently at the behest of Mexico's new president, José López Portillo) began releasing frequent and substantial updates of Reforma's probable petroleum reserves.

In April 1977, the director-general of Pemex, Jorge Díaz Serrano, was reported as stating that a lower-bound estimate of the sum of proved, probable, and potential commercially recoverable petroleum reserves in the Reforma Trend would exceed 60 billion barrels.[3] Moreover, large additional petroleum reserves were known to lie under the adjacent offshore waters of the Campeche Sound. At the time of Serrano's announcement, the Reforma Trend's first four onshore structures actually under development had yielded 11 billion barrels of proved reserves; initial drilling in Campeche's first three offshore structures revealed an additional 3 billion barrels of proved reserves. Furthermore, exploratory drilling had confirmed the presence of substantial oil in nearly a score of other structures that had yet to be developed, and there remained to be drilled more than 100 already-mapped structures lying within the geographical boundaries of Reforma's and Campeche's known productive trend. Exaggeration is common when cash-poor, overpopulated countries discuss their oil potential. Nevertheless, these known geological facts supported the inference that Pemex's April 1977 projections of southeastern Mexico's petroleum potential were indeed conservative. By late 1977, articles in the oil trade press were suggesting that even if additional exploration did not result in further extensions of the geographical boundaries of Mexico's new oil regions, their recoverable reserves were likely to be in the neighborhood of 100 billion barrels.[4] If additional drilling demonstrated that the Campeche trend continued farther offshore into the waters of the Gulf of Mexico, reserves in excess of 200 billion barrels were thought possible.

Pemex's subsequent exploratory drilling has proven successful. Thus, on September 1, 1978, President Portillo reported to the nation that Mexico's potential reserves of petroleum had now risen to 200 billion barrels.[5] At 60 billion barrels, the Reforma Trend already ranks as the world's third largest productive area—exceeded only by the legendary Persian Gulf giants of Bergen in Kuwait and Ghawar in Saudi Arabia. At 200-plus billion barrels of commercially recoverable petroleum reserves, Mexico would be catapulted into the position of being Saudi Arabia's co-equal.

Petroleum finds of the magnitude and location of Reforma's and Campeche's hold the potential for transforming the world's energy picture in the 1980s. Assuming Pemex maintains its present target of a reserve-to-annual production ratio of 20:1, reserves of 60 billion barrels could support sustained production of crude oil and associated natural gas at a daily rate totaling 8.5 million barrels of crude oil equivalents.* Higher reserves could support proportionately higher outputs. Since Mexico's domestic petroleum needs are likely to total only about 2 million barrels per day by the late 1980s, her export potential is enormous.

---

* The standard unit of measurement for natural gas is one thousand cubic feet (or Mcf). An approximate conversion factor is 7 Mcf. which has an energy content equivalent to one barrel of crude oil.

IMPORTANCE OF MEXICAN PETROLEUM FOR THE UNITED STATES

Long before the 1973 Organization of Arab Petroleum Exporting Countries (OAPEC) oil embargo, the United States was publicly committed to limiting its reliance on imported oil to levels that would not endanger national security in the event that the flow of oil imports was interrupted. Since the embargo, achieving this goal has been thought to require a sizable reduction from present oil import levels. However, known American reserves of crude oil and natural gas capable of being produced commercially by conventional means are inadequate to support sustained output expansions and are probably insufficient to halt the post-1970 decline more than temporarily. Therefore, in addition to embracing the poorly defined "motherhood" goal of much greater energy conservation through increased energy efficiency, two basic options have been advocated for meeting future energy demands of the United States without creating an unacceptable level of reliance on imported fuels. One option is to begin immediately to introduce the huge government-sponsored research and development (R & D), subsidies (and taxes) necessary to facilitate an environmentally acceptable massive switch from petroleum fuels to coal or nuclear power and to stimulate large-scale commercial production of nonconventional petroleum substitutes, such as coal synthetics, tar sands, and oil shale (which are presently uneconomical and, in most cases, technically unproven on a large scale). The second option is to introduce the huge R & D, subsidies, and taxes necessary both to sever the historic linkage between a growing gross national product (GNP) and growing energy consumption and to foster the rapid commercial development of renewable energy sources best exemplified by solar power.

Adoption of either energy strategy would be enormously expensive and time consuming. Moreover, even with substantial government incentives and encouragement, it is uncertain whether either strategy could succeed in reducing American reliance on insecure Organization of Petroleum Exporting Countries (OPEC) oil. A principal theme of this volume will be that the discovery of enormous amounts of crude oil and natural gas in southeastern Mexico raises the possibility that for at least ten years, the United States can simultaneously reduce its consumption of *insecure* oil imports and cut back sharply on the size and intensity of its commitment to perfecting either or both of the two basic energy options. But to do so will necessitate implementing policies that are aimed more directly at encouraging the rapid acceleration of the production and export of Mexican oil and natural gas.

DEVELOPING MEXICO'S PETROLEUM

The United States is a likely source of encouragement, and possibly extensive assistance, to further the accelerated production and export of Mexican petroleum. Because Mexico is now known to possess huge reserves of commercially recoverable petroleum, it needs to satisfy only two conditions before it can become one of the world's leading producers. First, the Mexican government must decide that rapid development and production of the nation's petroleum treasure is in the national interest. Second, Pemex requires access to sufficient financial resources and technical expertise to allow production to be built up rapidly to the much higher levels that can be sustained by Mexico's large and expanding petroleum base.

The government of President Portillo has already moved to satisfy the first condition. President Portillo recognizes that large exports of Mexican crude oil, natural gas, and their refined product and petrochemical derivatives offer the only credible means of allowing this nation of 64 million to escape an oppressive cycle of poverty, unemployment, and underemployment and to deal effectively with a soaring population and a foreign debt exceeding $20 billion. Assuming maintenance of Mexico's mid-1978 price for oil exports (about $13.35 per barrel fob [free on board]), its gross annual revenues from the export of 1 million barrels per day of crude oil would be nearly $5 billion, and its annual net profits would almost certainly be in excess of $4 billion, or, at Mexico's present population, about $60 per capita.*

Satisfying the second condition will prove to be much more difficult. The Reforma and Campeche reserves are huge and very productive—only 125 wells were needed to produce 700,000 barrels per day in late 1977. Neverthless, if Mexico is to achieve its full petroleum-producing potential as quickly as is technically possible, year-to-year gains in production would be about 1 million barrels per day by the early 1980s. To achieve this technically feasible rate of expansion would require a fourfold increase from 1977 drilling levels. Even more difficult would be the job of building quickly, and largely from scratch, the infrastructure necessary to permit such an acceleration in output: gathering lines and storage areas, new deepwater ports, huge units to separate associated

---

* The minimum costs of developing fields with the depth, relatively accessible location, and high productivity (about 5,000 barrels per well per day) of the Reforma fields is probably about $1 per barrel. Pemex is considered to be a model for a national oil company. However, its policy has been to employ superfluous staffing. Therefore, Pemex's actual costs could be substantially higher. The cost inflation induced by deliberate policies, such as overstaffing, are probably most accurately viewed as one way Mexico can choose to spend its oil profits. See Chapters 3 and 4 for elaboration.

natural gas and sulphur from crude oil, and pipelines to transport natural gas to the industrial cities of northern Mexico and, eventually, to the enormous gas-hungry pipeline network of the United States. The well-known problems in building the trans-Alaskan pipeline and developing the North Sea's oil reserves' suggest that a rapid and potentially much larger expansion of Mexico's petroleum output would stretch the capabilities of a consortium of the largest and best-managed international oil companies.

Pemex was created in 1938 to take over the Mexican assets of the just-nationalized U.S. and British oil companies. It was the non-Communist world's first integrated national oil company, and it has been a principal model for subsequent members of the genre. Though Pemex's senior management has a good reputation, the company faces four difficulties common to national oil companies in most developing countries. First, it is short of cash. However, this deficiency can be remedied immediately if Pemex agrees to make long term contracts for the sale of a relatively small fraction of Reforma's oil and natural gas. Second, in comparison to large private oil companies, Pemex is plagued by low labor productivity due to featherbedding labor practices in blue-collar jobs and an overstaffed (some U.S. publications suggest corrupt) middle management. Third, the combination of Pemex's revolutionary birth and restrictive labor contracts sharply constrains its employment of foreign expertise. Charges of "foreign exploitation" and "loss of economic sovereignty" are not easily refuted in Mexico. Fourth, in addition to the enormous job of developing the petroleum reserves of Reforma and Campeche, Pemex is also charged with developing a refining and petrochemical business capable of becoming a major force in world markets. If Reforma's crude oil and natural gas are to be developed at close to the maximum technically feasible rate, Pemex will have to devote nearly all its resources to the effort. Even so, extensive foreign assistance will be necessary. The United States is the most obvious source for providing this assistance.

The United States and Mexico share a common border, a democratic-humanitarian tradition, and, as a direct result of large numbers of legal and illegal Mexican immigrants, a strong and growing cultural affinity. These reasons suggest a special U.S. commitment to helping Mexico alleviate her economic problems by aiding her efforts to expand petroleum output. But there are even stronger and more selfish reasons why the United States should encourage greater Mexican oil production.

Growing reliance on imports of OPEC oil has made it much more difficult, if not impossible, for the United States to accomplish many of its political, strategic, and economic goals. Perhaps most important, a sharp increase in the vulnerability of the United States to an interruption in oil supplies has been a direct corollary of a rapidly rising oil import dependence. Unfortunately, the probability of such interruptions is not insignificant. Renewed armed hostilities with Israel could trigger another embargo by the members of OAPEC; conservative regimes in several Persian Gulf countries may not be immune to either debilitating civil wars or sudden coups by anti-U.S., pro-Communist revolutionaries; and key shipping lanes are vulnerable to blockage by terrorists or potentially hostile governments. In addition to these dramatic possibilities, growing dependence on oil imports has helped to precipitate a substantial deterioration of U.S. balance of payments and has placed control over U.S. and world energy prices in the hands of a cartel of oil-exporting countries. These two factors, in turn, have played a key role in precipitating the post-1973 worldwide economic instability and stagnation.

Proposing to attack these problems at their source, Presidents Nixon, Ford, and Carter each introduced national energy plans designed to reduce U.S. reliance on imported oil. Their plans differed in significant respects—in particular, President Carter's National Energy Plan is far more complicated and places more emphasis on conservation incentives and less on incentives to promote increased domestic production of crude oil and natural gas. Nevertheless, each administration has endorsed a massive U.S. commitment to the commercial development of alternatives to conventional petroleum fuels. But large-scale production of any of these petroleum alternatives raises severe environmental and/or public safety issues. With the exception of a shift from petroleum to either coal or nuclear energy, all appear likely to be much more costly than imported oil at present world prices. By adopting an energy policy that promotes the large-scale production and importation of Mexican oil and natural gas, the United States can reduce significantly both the size of its near-term (that is, 1980s) commitment to producing alternative fuels commercially and the problems posed by the importation of OPEC oil.

POTENTIAL U.S. MARKETS FOR MEXICO'S PETROLEUM

The United States would provide a natural and readily accessible market for Mexican oil and natural gas. Since crude oil is a highly fungible product, after making appropriate allowances for differences in economically important factors, such as sulphur content and specific gravity, the price that a barrel can fetch in any market is determined by the cost of the oil it replaces. Presently, the Persian Gulf countries, especially Saudi Arabia, are the marginal source of crude oil for all major world markets, and thus, in the absence of price controls, crude oil's price in any market is the Persian Gulf price plus freight.

The net wellhead value of any barrel of crude oil is equal to its price at the final market minus all prior production, transportation, and marketing costs. Because of its relative proximity, transport costs to the U.S. Gulf Coast are less than to all other potential major markets for Mexican oil exports. Conversely, the relatively long distance from the Persian Gulf means that oil fetches a higher price at the U.S. Gulf than in most other major world markets. The implication of these two facts is that if Pemex's goal is to maximize its net revenues or profits from the sale of its oil, then the U.S. Gulf Coast ports are the most desirable destination for Mexico's oil exports. Upon entering a U.S. Gulf port, Mexican oil can be transshipped easily anywhere in the United States east of the Rockies.* The potential market is enormous—this vast region was consuming about 7 million barrels per day of imported crude oil and refined products in early 1978.

The Reforma crude oil deposits contain unusually high concentrations of associated natural gas, which must be separated from the crude oil prior to sale. Presently, after being separated, large quantities of Reforma's natural gas are flared.† This wasteful practice can be halted only after pipeline factilities are put in place and domestic or foreign markets are developed. Some Pemex sources are already talking about the possibility of producing at least 10 trillion cubic feet of natural gas annually from the Reforma Trend by the mid-1980s.[6] Of this, approximately 5 trillion cubic feet (equivalent to annual crude oil shipments of 2.5 million barrels per day) would have to be either reinjected, flared, or exported because of insufficient domestic demand. Because liquefaction of natural gas and its subsequent transportation by tanker is many times more expensive than shipping natural gas via pipelines, the United States is the natural market for nearly all of Pemex's natural gas exports.‡

U.S. production of natural gas peaked at 22.5 trillion cubic feet in 1972. By 1977, production had fallen to 20 trillion cubic feet. Since crude oil products (especially fuel oils) are the closest substitute for natural gas and since U.S. current production of crude oil satisfies little more than half of current domestic demands, in 1978, U.S. oil imports were nearly 1

---

*Mexican oil could also be shipped to the U.S. West Coast. However, since the completion of the Alaskan pipeline, this region of the United States has actually had a substantial oil surplus, which is likely to continue until at least 1985.

†In many parts of the world associated natural gas is reinjected into the field in order to enhance ultimate oil recovery (by helping to maintain the field's pressure). Gas reinjection is not expected to be practiced widely in Reforma because waterflooding (that is, the injection of water) is thought to provide a much better method of pressure maintenance for fields of the Reforma/Campeche type and location.

‡There are unofficial estimates that it would cost about $.40 per Mcf to pipeline natural gas from Tabasco to the Texas border, whereas liquefaction would cost about $1.60 per Mcf.

million barrels per day higher than they would have been if natural gas production had not fallen over the preceding five years. Thus, substantial imports of Mexican natural gas definitely would result in a sharp reduction in U.S. oil imports.

Since the early 1950s, the gap between total domestic energy production and total domestic consumption has widened steadily in the United States—oil imports have risen sharply in order to make up nearly the entire shortfall in domestic energy supplies. Because most energy is used to fuel long-lived capital assets and consumer durables, there is a tight linkage between total energy demand and GNP, which can be modified substantially only over a period of several years. Similarly, the five to fifteen-plus year lead times necessary to develop potentially commercial, large-scale domestic energy projects require that changes from already planned sources of future energy supplies be gradual. In addition, because most domestic energy supplies are being produced at or near capacity, the United States has had little control over the amount of its total oil imports since the late 1960s, and it will continue to have little control over the next several years.* These facts imply that, from the late 1970s through the mid-1980s, the chief observable consequence of Mexico's rising exports of crude oil and natural gas will be to change the composition of U.S. energy imports. Mexico's share of the U.S. energy market will rise at the expense of a declining share from the OPEC countries, especially OPEC's Persian Gulf members. However, if the optimistic assessments of Reforma's and Campeche's petroleum reserves are proven correct (we should know by 1980 at the outside) and if the United States consequently chooses to delay or reduce the economic-political incentives needed to accelerate the commercial development of domestic alternatives to petroleum, then Mexican crude oil and natural gas could begin to be substituted for these other fuels by the late 1980s.

SECURITY IMPLICATIONS FOR THE UNITED STATES

Members of OPEC supply more than 90 percent of the non-Communist oil currently traded internationally. Because imported oil has no substitutes presently available at short notice, all of the large oil-importing nations recognize that sudden cutbacks in anticipated exports of OPEC oil can cause economic havoc. The cutbacks necessary to precipitate worldwide economic dislocations need not be large: the

---

* This statement assumes that the United States would not impose effective controls on oil imports. If such controls were to be imposed, the consequence would be either a rise in domestic oil prices or substantial shortages, which could only be alleviated by introducing rationing.

maximum cutback in world oil shipments during the nearly disastrous 1973–74 OAPEC embargo was only 4.5 million barrels per day.[7]

Future sudden cutbacks in world oil supplies will have one of two causes: a coalition of significant oil-exporting countries will decide to reduce sales in order to raise world oil prices or to persuade the oil-importing countries to accede to political demands; or violence will either disrupt oil production or interfere with its transport. The first type of interruption is likely to occur only as long as the oil exporters retain sufficient monopoly power to take joint action to restrict supply.[*] Violence is more likely to be a threat to those oil supplies produced in or shipped near countries facing potential domestic political turmoil or located in politically unstable regions.

Mexico is a developing country with a large, rapidly growing population and an enormous foreign debt. Unlike nearly all of the important oil-exporting countries, Mexico is not an OPEC member, and its large and impoverished population has a nearly insatiable demand for goods and services. These facts make it likely that as long as OPEC is able to maintain world oil prices near their present level, Mexico will continue to follow its present policy of being an expansionist oil producer. OPEC's December 1976 price split and its inability to agree to higher prices in December 1977 are widely attributed to increased competition brought about by recent rapid growth in oil production from three non-OPEC sources: the North Sea, Mexico, and Alaska. A gradual recognition by several leading OPEC producers that they would have to cut back their oil production in order to accomodate growing sales of Mexican oil would reduce substantially the probability of a future oil embargo.

In addition to Mexico's independence from OPEC, her geographical location greatly enhances the desirability of Reforma's petroleum. The natural gas pipelines between Tabasco and Texas should be approximately as secure as pipelines located wholly within the "lower 48" U.S. states. The roundtrip tanker voyage between the Bay of Campeche and Port Author, Texas takes about a week rather than the 90 days to and from the Persian Gulf. Also, from U.S. perspective, the international waters of the Gulf of Mexico are already among the most secure in the world. Moreover, a more than 50-year tradition of reasonably friendly relations between the two neighboring nations makes it unlikely that Mexico would opt to suddenly suspend contracts providing for on-going shipments of oil exports to U.S. importers. In short, imports of Mexican oil would be much more secure than imports of similar amounts of OPEC oil. Therefore,

---

[*] A monopolist gets its power to set a higher than competitive price from its ability to control output. But control over output is necessary for an effective embargo. Hence, sellers must have monopoly power in the world oil market if their embargo threats are to be credible.

U.S. oil security would be enhanced if development of Reforma's huge reserves makes it possible to substitute large amounts of Mexican oil and natural gas for OPEC oil, especially from the Persian Gulf. But the implication of this analysis is that if the United States can increase sharply Mexico's share of the total U.S. market for oil imports, then it is probably neither necessary nor desirable to adopt policies specifically aimed at reducing U.S. consumption of imported oil.

ECONOMIC BENEFITS TO THE UNITED STATES

After making allowance for the appropriate quality differentials, Mexican oil will sell in the United States at the same price as the OPEC oil it replaces. Therefore, except for·the likelihood that growing exports of Mexican petroleum will constrain OPEC's future pricing, U.S. consumers will not reap direct financial savings by a switch to this source of energy supplies. Nevertheless, the switch to Mexican oil and natural gas will yield four indirect, but valuable, economic or environmental benefits: a reduced deficit in the U.S. balance of payments; reduced requirements for high-risk, long-term capital investment; reduced need to subsidize the commercial development of presently uneconomic and, in most cases, as yet technically unproven alternative types of energy; and reduced environmental pollution. Sharply higher exports of Mexican petroleum also seem likely to yield a more intangible, but nevertheless valuable, dividend for both countries: by fostering faster economic growth within Mexico, there is likely to be some reduction in the rate of illegal Mexican immigration into the United States.

Unlike super-rich and sparsely populated countries, such as Saudi Arabia or Kuwait, most of the foreign exchange Mexico reaps from the export of its petroleum will be spent almost immediately in acquiring foreign goods and services. Moreover, because of its geographical proximity and comparative advantage in producing the agricultural commodities and industrial goods Mexico wants to import, the U.S. share of Mexico's total import market already is, and almost certainly will continue to be, far higher than its share of import markets of any non-Western hemispheric OPEC nation. Chapter 7 explains in more detail how both of these effects will work toward reducing the U.S. balance of trade deficit; therefore, it should help to strengthen the dollar.

Because of Reforma's enormous size and productivity, production of Mexico's petroleum can be accelerated sharply above presently planned levels within two to three years of a firm decision to do so. In contrast, with the exception of the natural gas at Prudhoe Bay on the Alaskan North Slope, the United States had, as of mid-1978, no large new commercial petroleum reserves capable of being developed rapidly. Unfortunately, construction of the trans-Canadian pipeline necessary to deliver Alaskan natural gas to the upper Midwest markets of the United States would take at least three times as long and cost seven to ten times as much as construction of a pipeline that would transport the same quantity of Mexican gas from the Reforma fields to the Texas border. Similarly, projects to develop the relatively abundant nonpetroleum energy sources of the United States—coal, oil shale, nuclear power, and coal synthetics— will take five to 15 years to complete and entail capital investments several times higher than those necessary to produce and deliver similar quantities of Mexican oil or natural gas to the United States. For these reasons, a decision to place more emphasis on developing Mexican oil and gas and less immediate emphasis on developing alternative domestic fuels would lead to a large reduction in requirements for high-risk, long-term capital investments.

Coal and nuclear power are economic alternatives to crude oil and natural gas at present prices. However, with present technology, use of both fuels is essentially restricted to heating large boilers.* For such uses as transportation, most small-scale heating (for example, homes), and industrial applications requiring clean combustion, there are presently no commercial substitutes for natural gas and refined oil products. The development of presently nonconventional substitutes for petroleum is thought to require massive public subsidies and to entail a high chance of technical and/or commercial failure. Greater reliance on Mexican oil and natural gas would allow more time to conduct the basic research that will help to determine which of the potential petroleum substitutes is most promising.

Another positive corollary of a U.S. decision to rely more heavily on imports of Mexican oil, while deemphasizing present commitments to the development of nonpetroleum energy forms, would be a significant reduction in future environmental costs. The production, transportation, and combustion of natural gas, crude oil, and their derivative products places far fewer demands on the environment and raises much less serious public health and safety issues than the production, transportation, and combustion of any of the alternative fuels likely to be commercially significant in the next 20 years.

CONCLUSION

Mexico and the United States have many issues to resolve before accelerated oil development and trade can become feasible. To recapitulate, Mexico has the potential to earn enormous profits from sharp

---

* Use of nuclear power is even more restricted than the use of coal. Its only commercial use is to generate electricity.

increases in its oil exports, which, in turn, could fuel its transformation into a far more prosperous economy. Because of the enormous size and productivity of its oil fields, the total cost of producing the great bulk of Campeche's and Reforma's crude oil are almost certainly less than $3 per barrel—a small fraction of the $13-plus this crude oil sells for on world markets.* The United States is the most likely market for the bulk of Mexico's petroleum exports. Sizable increases in U.S. imports of Mexican oil and natural gas would yield correspondingly large benefits: increasing the efficiency with which U.S. energy demands are satisfied, reducing OPEC's economic and political power, reducing the danger of serious interruptions to U.S. access to imported oil due to foreign wars or civil unrest, and lessening environmental degradation, by allowing the nation to scale down its commitment to developing alternative fuels.

In sum, since both Mexico and the United States would profit from a policy to expand production of Mexico's crude oil and natural gas at the highest technically feasible rate, the crucial question becomes: What policies are available to each country for promoting achievement of this goal?

Pemex's present plans call for the quadrupling of Mexico's crude oil exports to 1.1 million barrels per day in 1980 and raising natural gas exports from near zero to 2 billion cubic feet per day by 1982. Based on Pemex's record since the first Reforma discovery of always outperform-ing the announced plan, achievement of these targets seems likely. However, because the lead times for a significant increase in southeastern Mexico's petroleum output are two to four years onshore in Reforma and three to six years offshore in Campeche, a substantial speedup in meeting these targets is unlikely.

After 1982, Pemex will have more discretion in increasing the magnitude of yearly output at Reforma and Campeche. But given the unavoidable lead times, it must begin now to make the plans and investments necessary for much more rapid production. The principal constraints slowing down the rate of long-term expansion are current shortages of capital and technical expertise—resources of which the United States has an abundance. Since both countries would reap large gains from accelerating the rate of expansion, policies should be aimed at alleviating these constraints.

Pemex has already introduced two measures to circumvent its technical and financial shortcomings. First, a U.S. firm (Brown and Root, a Haliburton subsidiary) has been hired as project manager for the $500 million engineering and construction job necessary to develop initial production of 360,000 barrels per day from the Campeche Sound. This marks an unprecedented departure from Pemex's past policy. Since

---

* See Chapter 3 for elaboration.

nationalization, Pemex has nearly always employed in-house management expertise, purchasing only equipment for overseas.

Pemex's second move was to approach several U.S. natural gas transmission companies to help finance the 2 billion cubic feet per day natural gas pipeline from Tabasco to Texas by agreeing to prepay for the gas. Letters of intent were signed by six U.S. companies, who were willing to pay $2.60 per Mcf, and additional credits of $300 to $600 million were to come from the U.S. Export-Import Bank.

Mexico has displayed considerable flexibility and ingenuity in its efforts to promote the production and export of Reforma petroleum—to date, the U.S. government cannot make a similar claim. Displeased because the contracted delivered price for Mexican gas exceeded President Carter's proposed $1.75 per Mcf ceiling on new domestic natural gas, the Department of Energy refused to authorize the U.S. gas companies' purchases. Moreover, the Senate passed a resolution asking the Export-Import Bank to withhold advancing any financial credits until the pricing problem was resolved. In addition to demonstrating a remarkable disregard for Mexican sensibilities, these measures suggest that the U.S. government does not appreciate the important contribution Mexican petroleum can make toward satisfying the future energy demands of the United States.

Given the repeated assertions about the enormity of the U.S. energy problems, why is so little attention paid to promoting the option that has the most potential for alleviating them without imposing unacceptable economic or environmental costs? One possible reason is ignorance. But as knowledge of Mexico's enormous petroleum reserves rapidly accrues, this explanation becomes less plausible. At least as plausible is the explanation that the enormous potential of Mexican oil and natural gas has been downplayed precisely because admission of this fact would imply that the midterm (that is, 1985–95) problems of energy supply of the United States may be much less severe than many parties to the energy debate now wish to believe. Many participants in the debate over the direction of United States energy policy have a strong interest in insisting that the energy problem is enormous. Depending on their viewpoints as to the merits of economic growth, this premise has been used to justify either immediate massive subsidies and taxes to encourage production and consumption of petroleum alternatives or massive subsidies and taxes to encourage revolutionary life-style changes. The recent discovery of enormous, readily developable reserves of Mexican petroleum seriously undercuts both of these positions.

NOTES

1. "Oil Gives Mexicans a Boost, But They Plan to Stay Out of OPEC," *Wall Street Journal*, February 11, 1978, p. 1.

2. Alvaro Franco, "Mexico's Crude-Exporting Role May Be Short-Lived," *Oil and Gas Journal*, May 26, 1975, p. 25.

3. "Pemex Has New Chiapas-Tabasco Finds," *Oil and Gas Journal*, May 2, 1977, p. 120.

4. One of the better and more accessible technical assessments of southern Mexico's petroleum potential is A. A. Meyerhoff and A. E. L. Morris, "Central American Petroleum Potential Centered Mostly in Mexico," *Oil and Gas Journal*, October 17, 1977, pp. 104–09. Meyerhoff and Morris (p. 107) indicate that the "stratigraphic kinship of southern Mexico with the Chapayal basin suggests Guatemala should have valuable potential reserves."

Based on his discussions during a recent trip to Mexico and the assessment of a team of oil experts from his company, Armand Hammer, chairman of the board of the Occidental Petroleum Corporation, remarked at its annual meeting in May 1978 that southern Mexico could have petroleum reserves of 100 billion barrels. The *Oil and Gas Journal* reported on June 5, 1978, that the Mexican government had no quarrel with Hammer's assessment.

5. "Mexico's Oil, Foreign Exchange Reserves Soar; President Cites Economic Recovery," *Wall Street Journal*, September 5, 1978, p. 6.

6. Alvaro Franco, "Giant New Trend Balloons SE Mexico's Oil Potential," *Oil and Gas Journal*, September 19, 1977, p. 84.

7. "Aramco Sees Rapid Restoration of Crude," *Oil and Gas Journal*, January 28, 1974, p. 94.

I

SETTING THE SCENE:

FACTORS AFFECTING MEXICAN–U.S. ENERGY RELATIONS AND THE RATE OF DEVELOPMENT OF MEXICO'S RECENTLY DISCOVERED PETROLEUM

2

THE FIRST ERA
OF MEXICAN OIL:
1876–1938

Mexico is unique as an oil-producing region for two reasons. First, it has a long-standing oil economy—once controlled exclusively by foreign oil interests. Mexico provided nearly 25 percent of the world's supply of crude oil during World War I. Second, in 1938, Mexico became the second (after Bolivia) non-Communist state to nationalize the production of crude oil and natural gas through the expropriation of all foreign oil properties. An examination of Mexico's earlier oil experience is vital for understanding the nation's attitudes toward the production and export of its recently discovered petroleum treasure.

Mexico's first successful oil well was drilled in 1876. Thereafter, the course of the Mexican oil industry was to be set by the interaction of geological, economic, and political factors. Prior to 1910, the coupling of a favorable investment climate with large discoveries led to the establishment of a large crude oil industry on the Mexican Gulf coast. U.S. and British companies competed for dominance in both the crude oil export market and the domestic market for refined products. Between 1910 and 1920, the development of Mexico's oil industry was largely shaped by the chaotic events of the Mexican Revolution and by the demands to fuel a mechanized war in Europe, which included the rapid conversion of the U.S. and Royal navies from coal to oil propulsion. Though the turmoil accompanying the revolution was to force most nonoil foreign enterprises out of business, the U.S. and British oil companies enjoyed a huge growth in their sales of Mexican oil. This growth in the face of adversity helped to create the image of foreign oil companies as the financiers and manipula-

---

Leslie Fenlon, captain in the U.S. Navy, retired, co-authored the history between 1876 and 1920; Ernesto Molina, Jr., co-authored the history between 1920 and 1938. Dean Goodermote aided in researching this chapter.

tors of the revolution. World War I marked the high point of the first era of the Mexican oil industry. Thereafter, dwindling reserves, a hostile political climate, and most important, the discovery of larger reserves in other countries prompted the oil companies to reduce their investments in Mexican oil.

Throughout the entire period, events were to be influenced by two interacting but separable Mexican attitudes or positions: anti-Americanism and economic nationalism. Anti-Americanism had its roots in an earlier time, when it was the "manifest destiny" of the U.S. republic to consolidate its southern boundaries and bring Texas and California into the Union. In 1846, Mexico was invaded and defeated—it lost nearly half of its territory to the United States. Bitterness over these losses was to become part of the Mexican heritage, exacerbating future relations with the United States. Economic nationalism—a determination to prevent exploitation by foreign enterprises, especially the United States and its huge oil companies, and to control its own economic destiny—became a parallel concern (and it extends to modern Mexico's political life). Both attitudes took root during the early days of the Mexican oil experience.

THE MEXICAN OIL INDUSTRY UNDER DÍAZ: 1876–1911

The Climate for Foreign Investment

Porfirio Díaz ruled Mexico with an authoritarian, semi-Constitutional hand from 1876 to 1911. A self-made man, Díaz was born into poverty and advanced to the presidency after a successful army career. Díaz sought to raise Mexico from its poverty through industrial development; social advancement by industrial modernization was his creed. His regime became identified as the *científicos*, because he surrounded himself with the best scientific minds available. Under Díaz, serious and successful attempts were made to diversify the Mexican economy. Foreign capital investment in Mexico was encouraged by land and financial concessions, by tax abatements, and by establishing an atmosphere of order and stability through the president's personal control of the federal army, the city police, and the *rurales*—a strict, formidable paramilitary force that dominated the rural areas. Not only was there a stable atmosphere of law and order but wage scales for Mexican labor were extremely low, and unionization was nonexistent. Thus encouraged, foreign industrial and financial entrepreneurs moved into Mexico.

Between 1877 and 1901, massive roadbed construction extended Mexico's 417 miles of railroads to a length of 9,600 miles.[1] This expansion

was accomplished primarily by U.S. interests.* Generous Mexican subsidies, exceeding $100 million by 1901, encouraged the building of
railroads by such legendary U.S. railway barons as Collis P. Huntington,
Jay Gould, and E. H. Harriman.[2] Foreign mining companies arrived to
exploit rich deposits of gold, silver, copper, lead, and zinc. By 1908, U.S.
companies accounted for 840 of the nearly 1,000 foreign mining firms
operating in Mexico. The Guggenheim family alone operated 64 of these
mining companies.[3] By 1902, Mexico surpassed the United States as the
world's leading producer of silver, and in 1904, she became second to the
United States in copper production.[4]

Though important, U.S. enterprise did not dominate the Mexican
market. Indeed, prior to 1900, Britain's Mexican investments actually
outpaced those of the United States. The British concentrated their
investments in railroads, mining, land colonization projects, and oil. The
French invested primarily in government securities and banking, although the Boleo mine in Lower California—one of Mexico's richest
copper mines—was controlled by the French Rothschilds.[5] Germany
devoted its initial investments to trade, but by 1910, its investments
showed significant growth in all fields.[6]

The *científicos*, led by Díaz's secretary of finance, José Yves Limantour, brought about many financial reforms. A flexible and powerful
banking system was established, and in 1904, Mexico, the world's largest
silver producer, demonstrated its commitment to financial orthodoxy by
adopting the gold standard. As a result, Mexico was able to refinance its
national debt in 1911 at a 4 percent interest rate—far below prevailing
rates in the rest of Latin America.[7] In sum, by encouraging foreign
investment, Mexico reaped impressive industrial and commercial growth
during the first 25 years of the Díaz presidency. Nevertheless, around the
turn of the century, Díaz became concerned about the predominant
influence foreign capital, particularly that of the United States, was
exercising in Mexico. Whether this fear was instigated by the criticism of
his political opponents or by his own assessment of the economic situation
is hard to ascertain. In any event, in 1903, he moved to remedy the
problem by having Mexico begin to buy foreign-held railroad securities.
By 1909, Mexico had purchased control of most of the important railroads, and a new company, The National Railways of Mexico, was
organized, with the government holding a majority interest. The Mexican
railroad system had been effectively nationalized, and because the
interests of the remaining foreign stockholders were protected and

---

*The major British railroad enterprise in Mexico was the reconstruction by Weetman D.
Pearson of a railroad across the Isthmus of Tehauntepec, which linked the Pacific Ocean and
the Gulf of Mexico.

dividends continued to be paid, there were few complaints.[8] The discovery and development of Mexico's oil resources took place against this background of government-sponsored industrialization.

The Early Stages of the Mexican Oil Industry

Discoveries of oil seepages along Mexico's eastern coastal plain were recorded centuries ago in Aztec and Mayan histories. Stretching from the Texas border along the Gulf coast as far as Yucatan, these exudes marked the oil-bearing coastal plain of eastern Mexico. Heavily forested, with occasional cropland, and drenched by up to 100 inches of rain each year from across the Sierra Madre Oriental, the coastal plain trended south and then southeast from Tampico and the Panuco River. Along the entire way, the early names of various localities reflected the presence of petroleum exudes—El Chapopte, El Chapopotal, Chapopotilla, Cerro de la Pez, Ojo de Brea, all words meaning tar or pitch and of special interest to the oil prospector.[9] This coastal plain is where the early Mexican oil industry began. It is, also, at its southern extremity, the scene of the Reforma and Campeche fields (see Map 2.1).

Mexico's oil industry had an inauspicious start. In 1876, a Boston sea captain brought back a quantity of tar from Tuxpan, obtained financial backing, and returned to Mexico to drill for oil. He sank a few wells to a depth of about 500 feet, found a small amount of oil, and built a primitive refinery on an island in the Tuxpan River. He sold his product to the natives as illuminating oil. But his costs were prohibitively high, so when his Boston partners refused to advance additional financing, he committed suicide.[10]

The second attempt to produce oil from the Tuxpan region also proved unprofitable. England's Cecil Rhodes, looking for new fields to conquer, became interested in exploiting Mexico's oil potential. He formed the London Oil Trust to undertake investments in Mexican petroleum. After a substantial sum of money had been spent without a profitable return, certain of the holdings were sublet to another British company, the Mexican Oil Corporation. Again, the oil development effort was abandoned—not because there was no oil but because the company could not return sufficient profits.* It remained for three entrepreneurs—Henry Clay Pierce and Edward L. Doheny of the United States and Weetman Dickenson Pearson of Great Britain—to be the first to exploit the potential of Mexican oil successfully.

---

*The first British consular report on the potential of Mexican oil was prepared in 1887. George Jeffrey organized the Oil Fields of Mexico Company, Ltd., in 1898 to prospect in the Tampico area. Although he produced some oil, his wells were never productive enough to

Map 2.1: Mexico and the Lower United States

Source: Constructed by the author.

Pierce arrived in Mexico from St. Louis in 1885 as head of the Waters-Pierce Oil Company. Although Pierce owned a 35 percent interest in the company, Waters-Pierce was an affiliate of John D. Rockefeller's Standard Oil of New Jersey and was chartered to bring Standard's marketing techniques to Mexico.° By the end of 1885, Pierce had secured an oil-importing concession from President Díaz and had built up an extensive marketing organization. Because import duties were much lower on crude oil than on refined products, Pierce imported Pennsylvania crude, which he refined (mainly into kerosene) at Veracruz and Mexico City.[11] The arrangement proved satisfactory to both parties: President Díaz obtained substantial import duties, and Pierce, with a monopoly of the Mexican kerosene market, enjoyed substantial pricing power.

In addition to importing, refining, and marketing oil, Pierce had other irons in the fire. One of his most important ventures was to acquire controlling interest and become chairman of the board of the Mexican Central Railroad, which ran from El Paso, Texas, to Mexico City, with a branch line to Tampico.[12] In May 1900, on behalf of the Mexican Central

---

warrant paying dividends on company stock. He sold some oil to Pearson, before Pearson's great oil strikes at Potrero, in order to allow the Pearson enterprise to meet contractual obligations in Great Britain. However, Jeffrey's Fubero fields never fulfilled expectations.

The London Oil Trust's effort during this same period was confined to expending £90,000 on geological studies in Mexico. But Sir Boverton Redwood, a prominent British geologist, returned an adverse report on the prospective productivity of the lands, and the London Oil Trust divested itself from the Mexican oil business, subletting certain properties to the Jeffrey organization, which became the Mexican Oil Corporation in 1904. Ironically, Edward L. Doheny subsequently obtained properties at Cerro Viejo and Chapopte from the London Oil Trust and produced substantial amounts of oil. Doheny attributed British failure to recognize and exploit Mexican oil potential to the young and inexperienced geologists Redwood sent to Mexico. (See Alfred Tischendorf, *Great Britain and Mexico in the Era of Porfirio Díaz* [Durham, N.C.: Duke University Press, 1961], pp. 122, 123, and 138, and Pan American Petroleum and Transport Company, *Mexican Petroleum* [New York: PAP&TC, 1922], pp. 26, 28.)

°Ralph W. Hidy and Muriel E. Hidy, *Pioneering in Big Business, 1882–1911* (New York: Harper & Brothers, 1955), p. 258. This work is one of two volumes on the history of Standard Oil of New Jersey. The other volume is George Sweet Gibb and Evelyn H. Knowlton, *The Resurgent Years* (New York: Harper & Brothers, 1956). Both volumes admit that the Waters-Pierce Company was an affiliate of Standard Oil of New Jersey but maintain that the parent company could not control the aggressive business tactics of Pierce. These tactics, plus Pierce's extremely high profits, were a matter of considerable concern to the Domestic Trade Committee of Standard Oil from the 1880s onward. Pierce paid little attention to their suggestions and was considered to be an aggressive individualist, who had trouble working as a member of a team. For these reasons, the Waters-Pierce Company was vulnerable to attack as a monopoly, and in 1900, was ousted from the state of Texas. It was subsequently reincorporated in Missouri and regained access to Texas on the basis of a sworn statement that the company's connections with Standard Oil had been severed. An investigation in 1909 showed that Pierce's affadavit was false, and the Waters-Pierce Company was once again and finally expelled from Texas.

Railroad, Pierce invited Doheny, who had discovered the first oil in Los Angeles in 1892 and had already made a fortune, to prospect for Mexican crude oil. The Mexican Central Railroad hoped Doheny would find oil near their right-of-way and as an inducement promised to purchase his initial production for use as locomotive fuel. Arriving at an area of oil exudes 35 miles west of Tampico, Doheny was enthusiastic:

> We found a small conical-shaped hill—where bubbled a spring of oil, the sight of which caused us to forget all about the dreaded climate—its hot, humid atmosphere, its apparently incessant rains, those jungle pests the *pinolillas* and *garrapatas* (wood ticks), the dense forest jungle which seems to grow up as fast as cut down, its great distance from any center that we would call civilization and still greater distance from a source of supplies of oil well materials—all were forgotten in the joy of discovery with which we contemplated this little hill from whose base flowed oil in various directions. We felt that we knew, and we did know, that we were in an oil region which would produce in unlimited quantities that for which the world had the greatest need—oil fuel.[13]

Doheny immediately commenced to purchase 450,000 acres in the area surrounding this and other exudes. Unknown to him, this purchase of surface land to obtain the subsoil mineral rights—a policy authorized by one of Díaz's first legislative achievements, the Mineral Act of 1884—was to create great problems in the future, because after the Mexican Revolution, Mexico reinstated the Spanish legal principle that subsoil rights can only be held by the nation. Doheny successfully obtained concessions for duty-free imports of all equipment necessary for producing crude oil, along with exemptions from all taxation, except the stamp tax, for ten years. He commenced drilling on May 1, 1900, at the exact spot where he had first seen the hill of oil, Cerro de la Pez. Early on the morning of May 14, at a depth of only 545 feet, oil rushed into Doheny's first well in such force that it lifted the tools off the bottom and interrupted drilling. Doheny had struck the Ebano oil field—the dawn of a commercial Mexican oil industry had begun. Doheny proceeded to drill additional successful wells in the Ebano field and to build storage tanks and pipelines. Within a short period, they also began production in the Casiano field, which was 65 miles southwest of Tampico.

Despite his success in finding and producing oil, Doheny faced many difficulties. Initially influenced by the failure of the London Oil Trust, the científicos were skeptical about the prospects for success of any oil enterprise. Additional investment funds proved difficult to obtain. As Doheny described the situation:

> Our plan was looked upon with little faith by most of the authorities in Mexico City. It was regarded as merely a Yankee scheme for selling oil

> stock in a plausible but unstable enterprise. Our undertaking to develop
> oil in reality received no encouragement except from the President of
> the Republic, Porfirio Díaz, and the Minister of *Fomento*. . . . From both
> of these honorable men we received assurances of friendship and
> encouragement that alone gave us the heart to proceed with this
> enterprise.[14]

Again, during the development of the Casiano fields, President Díaz
proved helpful. A major oil strike was imminent, but a recalcitrant
landowner was impeding completion of the pipeline to Tampico by
refusing to sell a right-of-way across his land. President Díaz solved this
problem by issuing a permit to Doheny to construct his pipeline across the
forbidden area.[15]

A second set of problems arose because the Mexican Central Railroad backed out of the purchase arrangement that had brought Doheny
to Mexico in the first place.° With oil filling his storage tanks, Doheny had
no market, and until Standard Oil of New Jersey came to his rescue, he
was forced to use his heavier oil for asphalting streets in Mexico City. As
Doheny's first large customer, Standard agreed to take 2 million barrels of
oil annually for five years and paid a large advance.[16] Eventually, a large
market for crude oil from Doheny's Mexican Petroleum Company
(MEXPET) was established in the New England states and other North
Atlantic ports; sales grew to 20 million barrles in 1921.[17]

By 1905, Doheny's production success and his supplier relationship
with Standard Oil was causing second thoughts among the cientificos.
Waters-Pierce, a Standard Oil affiliate, already had tight control over the
consumer marketplace in Mexico. Doheny testified at a later date:

> At that time the Díaz Government was very much opposed to monopo-
> lies, and General Díaz asked me point blank if we were in any way
> connected with the Standard Oil Co. When I told him no, he asked me
> to promise that I would never sell out to them without first letting him
> know, so that the Mexican Government could have the opportunity of
> buying the property before allowing it to pass into the hands of a very
> strong foreign organization.[18]

This fear of complete domination of the Mexican oil industry by a
U.S. monopoly prompted the entry into Mexico of the third great oil
entrepreneur, England's Weetman Dickinson Pearson, knighted in 1911 as
the first Viscount Cowdray. The choice of Pearson to diversify Mexico's

---

°It seems reasonable to infer that Pierce, in his capacity as owner of the Mexican
Central Railroad, canceled the purchase arrangement which had been made in 1900 because
he began to view Doheny as a potential competitor.

oil economy was logical. His contracting firm—S. Pearson and Son, Ltd.—was world famous for constructing the Dover harbor works and the Thames tunnel in England, as well as the Hudson River tunnels in the United States. His construction record in Mexico was equally prestigious and had resulted in a close personal relationship with President Díaz. Pearson's firm had built the Grand Canal to drain the Valley of Mexico; it had installed drainage, waterworks, electric lights, tramways, and a modern harbor at Veracruz; and it had reconstructed the Tehuantepec National Railroad across the Isthmus, building a harbor terminal works at either end. This latter enterprise best illustrates Pearson's close relationship with Díaz. Pearson provided half the capital for the government-organized Tehuantepec Railroad Company and managed all construction and repair. In return, he was promised 37 percent of the railroad's net profits for five years and a larger percentage thereafter. On the opening day, Díaz and Pearson rode with their sons the entire 170-mile length of the line.[19]

In 1901, Pearson began to prospect for oil near San Cristobal on the Isthmus of Tehauntepec. Planning to establish Mexico's first fully integrated oil company, he constructed a refinery, storage center, and pipeline at Minatitlan to handle the production expected from his San Cristobal wells. When San Cristobal's wells proved to be only modestly successful, he was forced to buy crude oil from other producers.

In 1906, the Mexican government granted Pearson vast concessions for oil on unoccupied government-owned lands in five states. Mexican Secretary of Finance Limantour approved the concession to prevent Standard Oil domination.[20] With products from the Minatitlan refinery and the hope of crude oil from his new concessions, Pearson mounted a serious effort to break the Waters-Pierce kerosene monopoly by obtaining a much larger share of the Mexican market. Pearson was challenging both American companies. A price war ensued with Waters-Pierce. The bitter fight was referred to at the time as the "great Mexican oil war." By 1910, with 40 railway cars hauling kerosene to various parts of Mexico, Pearson had garnered almost 25 percent of the retail trade.[21] Prices tumbled: the cost of a 20-liter can of kerosene fell from $3.50 in 1890 to only $.80 a year and a half later. As prices plummeted, Pearson absorbed severe losses. Waters-Pierce, with its long-established and effective marketing organization, was less seriously hurt. In 1910, the price war ended; prices were raised and stabilized at $2 for a 20-liter can.[22]

Pearson's search for oil in his new concessions was finally successful in 1910. After spending several million pounds sterling and losing his first oil strike to fire, Pearson struck the largest gusher known to the world at that time, Potrero del Llano No. 4, which yielded more than 100 million barrels of oil in just eight years. To develop this find Pearson formed the Mexican Eagle Oil Company, popularly known as El Aguila. After the incorporation of Mexican Eagle—Porfirio Díaz, Jr., was a prominent member of the board of directors—Pearson had no further financial worries in Mexico.[23] J. Fred Rippy declared in his analysis of British investments in Latin America that "Pearson probably garnered larger profits from Mexico than any other man, either during or since the Spanish Conquest."[24] The magnitude of the early Mexican oil discoveries is not generally well known, but an appreciation of its size and world impact may be gleaned from a brief examination of three great gushers.

On July 4, 1908, Pearson's drillers struck oil in the Dos Bocas well on the shore of the Laguna de Tamiahua halfway between Tuxpan and Tampico. The pressure in the well was so great that the derrick and four-inch drill pipe were destroyed completely. For unknown reasons, the oil ignited, and a 1,000-foot column of flame shot out of the well and burned uncontrollably for 58 days. When the raging fire was finally brought under control, the well produced only salt water. This spectacular blaze was reported worldwide and focused attention on Mexican oil.*

In December 1910, Pearson's drillers struck oil again in Potrero del Llano No. 4, a well located about 20 miles west of Tuxpan. Again, the pressure was tremendous, but this time, there was no fire. Oil flow was estimated at 100,000 barrels per day, and the well was capped only after 60 days of strenuous effort.

The supreme Mexican gusher was Doheny's Cerro Azul No. 4, a well located only ten miles north of Pearson's Potrero del Llano discovery. Doheny's crew struck a gas pocket during the night of February 9, 1916. The following morning, there was a huge explosion, which shot the drilling tools into the air. Seven hours later, the oil came—constantly increasing in volume and pressure until the thick column soared to a height estimated at 600 feet. For nine days, the oil gushed unrestrained, reaching an estimated maximum flow of 260,000 barrels in the 24 hours before the well was capped on February 19.[25]

As a result of these spectacular discoveries, Mexican oil was held in awe by the outside world in the first two decades of the twentieth century—in a way similar to the awe accorded the great oil fields of the Persian Gulf today. The oil strikes were so spectacular that even failures created a legend. A unique feature of Mexican oil was that gusher conditions eliminated the need for extensive pumping; most wells contin-

---

*It can be conjectured that the fires under the boilers, which provided steam to the drilling rigs, were not extinguished in time and that the oil drenched the boilers and was ignited. At the time, this was the recognized danger point in drilling for oil. In the case of Potrero No. 4, Pearson's drillers, having learned a lesson in the Dos Bocas catastrophe, acted with great presence of mind when the gusher commenced. Scrambling in the darkness through a thicket, which was the only way left open to the boilers, they put out the oil burners, thus averting another major conflagration.

ued steady production under their own hydrostatic and gas pressure. Furthermore, Mexican wells had unusually high average yields. Production from the 300 wells in Mexico averaged 1,800 barrels per day; in comparison, the 250,000 wells producing in the United States at the same time averaged less than five barrels per day.*

Unfortunately, the average Mexican, buffeted by poverty and social inequities, enjoyed few of the fruits of this oil bonanza. Therefore, the oil boom was widely viewed as another example of foreign economic domination supported and protected by Díaz and the científicos, who were perceived as the only real beneficiaries of the oil wealth. The Díaz regime failed to recognize that economic development in Mexico was merely placing an affluent industrial superstructure atop a poverty-stricken agrarian society, further widening the gap between rich and poor. In fact, the very nature of the Díaz dictatorship acted against the welfare of the rural Mexican. More and more land, much of it in common ownership of the village communities, was appropriated and enclosed by plantation owners or foreign corporations. By 1910, a mere 834 men owned one-fourth of Mexico's land.[26] Peonage was common and deliberate—the rurales maintained tight control; thousands died in penal settlements or while toiling on sisal plantations in the Yucatan. Despite outward appearances of progress, order, and stability, Mexico seethed and suffered.

For many Mexicans, the científicos came to symbolize all that was corrupt in society—not only a basic corruption of ideas but a more visible semblance of personal corruption, which was flaunted daily, as they indulged in the great wealth reaped through modernizing Mexico. Moving from one official position to another, including the governorships of the various states, the científicos grew wealthy in direct proportion to the increasing wealth of the industrial economy. There was little need for illegal sources of income. The vast process of modernization provided ample opportunity for personal investment and financial gain.[27]

By 1910, there was a swelling rumble of discontent throughout the country, not unlike the rumbling gas pockets that precede an oil gusher. The era of President Díaz was drawing to a close. The favors Díaz had granted the European capitalists—especially Pearson—aroused anger and resentment among Mexico's American community.[28] Standard Oil of New Jersey, with its wealth and political connections, was a formidable opponent for the aging president. This combination of popular discontent and U.S. business dissatisfaction was destined to immerse the Díaz

---

*This type of statistical comparison can be misleading. For example, the top 300 producing wells could be used rather than the total of 250,000, many of which were worn-out. (See Pan American Petroleum and Transport Company, [New York: PAP&TC, 1922], *Mexican Petroleum* pp. 96–108.)

regime in revolution, with foreign oil interests playing a prominent part in the revolutionary intrigue.

FOREIGN OIL INTERESTS AND THE MEXICAN REVOLUTION: 1910-20

The Mexican Revolution was started by members of a northern elite, the *Maderistas*, and became a conflict in which one elite opposed others. While initially an attempt to effect political change in the Díaz administration, the revolution was transformed over the years into a social restructuring of the Mexican society; it veered from political change to agrarian reform and developed strong antiforeign, prolabor tendencies, which included the expropriation of foreign oil properties. Though the revolution was not fully consummated until 1940, only its initial stages—from 1910 to 1920—were characterized by violence and bloodshed.

Of particular importance during the initial stages of the revolution was the fact that regime after regime acted too slowly in affecting the social and economic reforms that were being demanded. Consequently, insurgencies continued. Chaotic uprisings among the Indians, the somewhat more organized operations of "Pancho" Villa and Emiliano Zapata, and the increasing popular support for the northern aristocrat, Francisco Madero, forced Díaz to resign on May 25, 1911, and flee to France. Two weeks later, Madero entered Mexico City.

Madero attempted many important political reforms, but he failed to produce any real changes in the social order. In March 1912, Pascual Orozco and his army revolted. Although this revolt was put down by General Victoriano Huerta, it became clear to many that Madero could not reestablish order. When General Felix Díaz, the nephew of Porfirio Díaz, laid siege on Mexico City, Huerta defected, arrested Madero (who was later shot), and proclaimed himself president.

Huerta's corrupt and forceful rule was also ill-fated. Aided by President Woodrow Wilson's lifting of an arms embargo, the rebel armies of Zapata, Villa, and Álvaro Obregón gained control of the northern and central provinces of Mexico. This strange and unstable alliance, known as the Constitutionalists, fell under the leadership of Venustiano Carranza and was able to come to power after U.S. naval forces siezed Veracruz. Huerta resigned, and after a brief struggle, Carranza assumed the Mexican presidency.

Carranza introduced much needed social reforms. The hallowed Constitution of 1917, with its collectivist principles, was drawn up and implemented under his auspices. Yet, he too was destined to a fate like Madero's. Throughout the period, Zapata was in general rebellion against any regime that did not implement radical agrarian reform, and Villa pursued a course of banditry nearly devoid of ideological overtones.

Carranza's involvement in Zapata's assassination and his attempt to dictate his successor led to his downfall in 1920. He was hunted down and murdered. It was left to Obregón to consolidate the revolution.

In July 1923, Villa was killed by a relative of one of his victims. In ten years time, four of the five heroes of the Mexican Revolution—Madero, Zapata, Carranza, and Villa—met violent deaths. Obregón would follow. He was assassinated by a religious fanatic in 1928. Throughout the period, a bankrupt central government and rebel troops competed for financial assistance from the wealthiest sector of the Mexican economy—oil. Foreign oil interests found it expedient to grant or deny forced loans in accordance with their special interests.

The Fall of Díaz

José Vasconcelos, a long-time Madero advocate and, for a short time, one of his personal secretaries, described the atmosphere in Mexico in 1911 as follows:

> The end of the Porfirio Díaz regime marks the climax of American influence in Mexico. The large American concerns had become accomplices and associates of the Díaz functionaires in the adjudication of immense country estates, and in the ownership of mines, oil wells and industries. The American investor, the capitalist, was decidedly and constantly a warm adherent of Díaz and looked upon the commencing revolution as an outbreak of banditry that should be suppressed.[29]

The views of Vasconcelos concerning American business advocacy of Díaz were oversimplified. The positions of the major Mexican oil companies for or against Díaz were clearly defined by 1911. The English, led by Pearson's El Aguila company, clearly held a highly favored position in the Mexican oil market. Not only had El Aguila become the largest oil producer in Mexico, surpassing Doheny with his Potrero strike, but it had broken the Waters-Pierce monopoly of the retail oil market. In all of this, Pearson had been warmly and openly supported by both President Díaz and his finance minister, Limantour.

Pierce, on the other hand, was outraged by the preferential treatment being accorded the British. With his refinery dependent upon U.S. crude oil imports, Pierce demanded that the Mexican government remove the crude oil tariffs, which hindered his fight against Pearson, who had his own domestic crude oil production, thus avoiding the tariff and, thereby, underselling his competitor.* Furthermore, Limantour's effective nation-

---

*Pierce and Pearson were to settle their differences in 1913 by agreeing to divide Mexico's retail market.

alization of the Mexican railways in 1909 included the Mexican Central, previously controlled by Pierce. Although Pierce remained a major stockholder, a new board of directors, which was immune to his influence, was appointed. The new management imposed freight charges that discriminated against Waters-Pierce in favor of El Aguila.[30] As a result of the loss of control of the Mexican Central Railroad and the refusal of the government to raise the import tax, Pierce became a dedicated opponent of the Díaz regime.

Doheny and his Mexican Petroleum Company held a more ambiguous position at the start of the Mexican Revolution. Doheny's company had been very successful at the Ebano and Casiano fields and was not heavily pressed by Pearson, who had suffered several years of failure and had spent at least £5 million before gaining success at prospecting in 1910. Pearson's crude oil successes were still very recent in 1911 and certainly did not imperil Doheny's export market. Nor was Doheny in competition for the Mexican retail trade. His need for official sanction and assistance from President Díaz was a matter of past history. Despite this relatively stable existence, when, during the opening stages of the revolution, it was rumored that the United States might intervene on behalf of Díaz, Doheny cabled President Taft opposing intervention and stating that the British would be the only beneficiaries of such action.[31] It seems clear that from the U.S. perspective, the Mexican embroglio was becoming as much an Anglo-American commercial battle as an engagement between opposing Mexican forces.

In addition to Pierce and Doheny, there were several major U.S. oil companies—Texaco, Gulf, and, especially, Standard Oil of New Jersey—which though not engaged directly in Mexico were anxious to obtain their share of the oil wealth discovered by Doheny and Pearson. However, in the fading days of the Díaz regime, U.S. companies found it impossible to obtain Mexican concessions. It was always conceivable that a new Mexican government might be more amenable to granting new oil concessions, particularly if financial aid were made available to assist in the overthrow of Díaz.

How much oil money, if any, the U.S. companies may have invested in the Madero insurgency has never been revealed. There was testimony, shrouded in conjecture and controversy, that either the Waters-Pierce Oil Company or Standard Oil, and perhaps both, were involved in financing Madero. The companies issued vehement denials. However, these allegations had enough substance to prompt a subcommittee of the U.S. Senate Foreign Relations Committee to conduct hearings in 1912 on the subject of U.S. participation in the Mexican disorders.

Chief among these allegations was the saga of C. R. Troxel. In April 1911, claiming to be an agent for Standard Oil, Troxel made contact in El Paso, Texas, with revolutionary agents and allegedly offered $500,000 to

$1 million on condition that they issue his company a commercial concession once they became established in power. Troxel claimed to have a letter signed by John D. Archbold, acting head of Standard Oil, authorizing him to make contracts. Unbeknownst to Troxel, one of his El Paso contacts was an informant for the U.S. Justice Department. All negotiations with the Madero representatives were duly reported to the U.S. attorney general, who, in turn, passed the information to Secretary of State Philander Knox. After discussing the matter with President Taft, Knox dispatched a letter to Archbold, informing him of the serious and conclusive nature of the Justice Department's information and warning him that a vigorous investigation would be instituted to determine if Standard Oil had violated the neutrality statutes. Archbold denied his company's involvement and disclaimed Troxel as a Standard agent; coincidentally, his reply was dated May 15, 1911, the very day that the Supreme Court decreed the dissolution of Standard Oil, thereby, effectively terminating its special relationship with Waters-Pierce.[32] In any event, the eventual success of the Madero insurgency was signaled on May 10 when Ciudad Juárez fell to the Maderista forces, and this particular loan was never consummated.

The Senate investigating subcommittee also reviewed more circumstantial evidence implicating Standard Oil in financing Madero. Lawrence F. Converse, an American soldier of fortune with the Maderista troops, stated that Madero told him the Maderista forces were getting financial support from the Standard Oil Company, which would "back them to the last ditch."[33] Juan Pedro Didapp, writer, publicist, and sometime member of the Mexican diplomatic service, testified concerning conversations he had with Sherburne G. Hopkins, a Washington lawyer long active in Latin American revolutionary causes and Madero's legal adviser. In a series of meetings held in Washington, D.C., during 1910, Hopkins allegedly told Didapp that he would have no difficulty in obtaining the help of Standard Oil to overthrow Díaz.[34] Hopkins denied any actual financing of the revolution by Standard Oil when he testified before the subcommittee. However, it is interesting to note that as soon as Díaz capitulated, Hopkins went to work for the Waters-Pierce Company.[35]

The Senate investigating subcommittee had not completed its deliberations before Madero was deposed. The committee disbanded without issuing a final report. One member, however, proclaimed Standard Oil innocent of all charges. He was Senator Albert B. Fall from New Mexico, who, along with Doheny, would be involved in the Teapot Dome scandal a decade later.

Other evidence, again mainly circumstantial, implicates Pierce in funding the revolutionary cause. Hawley Copeland, a private secretary in the Waters-Pierce organization, wrote to President Wilson in 1913 that there had been a retired manager retained on the payroll whose wife was on such friendly terms with the Madero family that it could be "assumed" that she was the go-between in financing the revolution with either Waters-Pierce or Standard Oil money.[36] Percy N. Furber, an influential resident of Mexico and a former president of Oil Fields of Mexico, Ltd., told the financial journalist Clarence Barron that "H. Clay Pierce put up the money behind Francisco Madero and started the revolution."[37] Finally, Porfirio Díaz, Jr., attributed his father's demise to Pierce.[38]

The oil entrepreneurs made no confessions or admissions of complicity in the revolutionary events of 1910 and 1911. Nevertheless, it remains a fact that financial support of the *insurrectos* was the great oil issue of the first phase of the Mexican Revolution. Whether innocent or guilty, the oil companies had both the capability and motivation, while the Maderista forces had the need. Despite the claim of Vasconcelos that Madero never made any deals with foreign private interests to modify the existing regime of crude oil concessions, it is a fact that during the 1911–12 Madero era, Texaco made its first investments in Mexican oil; Gulf made its first foreign investment of any type by entering the Mexican oil industry; and the Magnolia Oil Company, controlled by the presidents of Standard Oil of New York and Standard Oil of New Jersey, purchased 400 acres of land in Tampico.[39] Whether or not Madero actually accepted foreign financial backing during the revolution, there was a wide perception—fostered in England and Mexico and widely reported in the press in the United States—that he was the recipient of oil money. This perception by itself was an influencing factor in attitudes toward the Madero regime and in the regime's own conduct and self-perception.

The Madero Regime

The actions of foreign oil interests during the Madero regime were characterized by a reversal of the roles of U.S. and British interests and by the aggressive intervention in Mexico's internal affairs by the U.S. ambassador, Henry Lane Wilson.

The fall of Díaz was a heavy blow to Pearson and the El Aguila oil interests. According to his biographer, J. A. Spender, Pearson

> found himself in a sea of political troubles, which threatened all his interests in Mexico, and most of all his oil interests. As the friend of Díaz and his partner in the economic development of the country, he was not likely to be in favor with the newcomers, and there was the further difficulty that none of them for the next few years had any firm grip on the country—all the governments proved to be transient and embarrassed phantoms fighting desperately for their lives with revolutionaries who found it profitable to make trouble by attacking the foreigners and endeavouring to hold them to ransom. The foreigner, and especially if

he were an Englishman, was in a peculiarly difficult position. He might call upon his Government to protect him, but, if he did so, he ran the risk of raising the Mexican people against him; and his own Government, moreover, had always to consider the delicate question of the Monroe Doctrine and the interpretation which might be put upon it at Washington. With an American President anxious to stay out of Mexico, and powerful interests in the United States unwilling that Great Britain should assume police duties in that country, the Englishman who looked to either to protect him was very likely to fall between two stools.[40]

Soon after Madero's triumphant entry into Mexico City, Pearson sought and received an interview. On August 26, 1911, he conferred with Madero, who gave him assurances that all concessions granted by the previous government would be respected and that additional foreign capital would be encouraged to come into Mexico. During the course of this conversation, Madero expressed his fear of Standard Oil domination in almost the same terms as in the Díaz-Doheny exchange in 1905: "They [the Madero regime] would look with suspicion . . . on the entrance of S[tandard] O[il] into Mexico, and he hoped therefore we would not sell out to these people."[41] When adverse reports of this meeting appeared in the Mexican press—planted, perhaps, by Waters-Pierce agents and claiming that Pearson's concessions had received no guarantees—Madero responded with a public declaration:

> Much has been said lately in the foreign press and has frequently been copied in our own to the effect that the new Government has the intention of making an investigation into the petroleum concessions granted to the house of Pearson and that this investigation has as its goal the revocation or restriction of these concessions. In this context I wish to state now, once and for all, that it has never entered into my plans to do such a thing, since I know that there is nothing irregular in the Pearson concessions and that those concessions were granted legally. . . .
> I have said to Mr. Pearson that I wish him to continue in this country his petroleum and other businesses, since he has brought here a great quantity of capital and destroyed the petroleum monopoly which formerly existed in the republic.[42]

Probably in the interest of consolidating these assurances, in March 1912, Pearson used his influence in Washington, D.C., to campaign for stricter enforcement of the prohibition on U.S. munitions exports to the forces of General Orozco, the leader of a rebellion against Madero in the northern provinces. Pearson's Washington legal counsel was Henry Taft, the President's brother. At Pearson's request, Taft discussed the problem with the president and the attorney general. Within eight days, a ban on munitions exports had gone into effect, and Madero's secretary of finance cabled thanks to Pearson for his cooperation.[43]

There is little doubt that despite these assurances, Pearson was nervous about Madero and, because of his previous close personal relationship with Díaz and an economic interest in restoring domestic tranquility, was regarded as a proponent of the Díaz faction in revolutionary politics. There is, however, no firm evidence that Pearson made financial contributions to the Porfirista cause. Once again, we are subject to perceptions colored by the emotions of the time, but perceptions that nevertheless have colored the legacy of the revolution. Several historians have accused Pearson of financing Madero's downfall.* Nevertheless, responsibility for this outcome would appear to be more properly laid at the feet of the U.S. ambassador to Mexico, Henry Lane Wilson.

Despite initial optimism that Madero could stabilize conditions in Mexico, Ambassador Wilson and the American community soon came to realize that he could not prevent attempted coups nor put down the counterrevolutionary movement of General Orozco. In the south, Zapata, though disarmed on three separate occasions, continued his agrarian revolt. Disillusionment set in the international community as depredations against foreign property owners and commercial concession continued. As foreign claims against the Modero government mounted, Ambassador Wilson took the offensive.

In early September 1912, the State Department provided Ambassador Wilson with the text of a note to be delivered to the Madero government over the ambassador's signature. The note amounted to a virtual ultimatum. The Mexican government was charged with laxity in protecting the interests of foreigners, particularly U.S. citizens, against the tyranny of petty local authorities and from anti-U.S. sentiment. Among other examples, the note specifically cited that U.S. oil interests in the vicinity of Tampico were being "taxed almost beyond endurance." This "predatory persecution, amounting practically to confiscation" must cease immediately, or the United States would no longer forbid the

---

*Cleona Lewis, in *America's Stake in International Investments* (Washington, D.C.: Brookings Institution, 1938), p. 222, states categorically that British oil money financed Madero's overthrow. The same contention is supported by Guy Denny, *We Fight for Oil* (New York: Alfred A. Knopf, 1930), p. 46, on the basis of Pearson's subscription to the Huerta counterrevolutionary "loan." The newspaper, the New York *World*, in its October 21, 1913, edition, openly attacked Pearson, accusing him directly of having supported the Orozco insurgency against Madero and implying that Pearson had aided in the overthrow of Madero. In response to numerous public accusations, Pearson emphatically denied having interfered in Mexican politics during the Madero regime or of having taken part in the overthrow of the Madero government, which, he said, had specifically recognized the validity of his concessions. This latter statement is true. For purposes of perspective, these denunciations of Pearson occurred at a time when Anglo-American commercial competition in Mexico was a highly emotional issue in the United States. See, also, the Cowdray papers, as quoted in Calvert, op. cit., pp. 237, 277.

exportation of arms to the rebels. The government of Mexico must either restore law and order or admit its inability to do so. In the latter case, the United States would be forced to "consider what measures it should adopt to meet the situation." [44]

Mexico's reply, by Minister of External Affairs Pedro Lascurain, was not delivered until mid-November. Lascurain denied that the new oil taxes had been levied out of anti-U.S. sentiment. Rather, the income from the $.03 per barrel tax was needed to run the government and establish peace. Furthermore, it was not a discriminatory tax: all producers, whether British or American, had received equal treatment. Subsequent to this exchange, Lascurain made an urgent trip to Washington, D.C., where, in high level discussions, both President Taft and Secretary of State Knox sought to impress him "that Mexico must protect American life and property; do justice to American citizens; restore order; and respond to the great moral obligation to be especially considerate of American interest." [45]

The September note, citing the possibility of military intervention to protect the Mexican interests of the United States, was a sudden and dramatic demarche in Mexican-American relations. It definitely marked a turning point for the Madero regime, whose next five months were characterized by a steady deterioration. Although the Taft administration did not follow up on its ultimatum, the note provided the independent-minded Ambassador Wilson with all the justification he needed to deploy his own plan for getting rid of Madero. Wilson pressed U.S. demands for claims upon a bankrupt government, threatened armed intervention, denounced the regime as a wicked despotism, and repeatedly urged Madero to resign. By mid-February 1913, the forces of Felix Díaz were fighting in Mexico City. Ambassador Wilson took three courses of action. First, he sought U.S. warships and marines for dispersal along Mexico's Atlantic and Pacific coasts, to be employed at his discretion in case of a crisis, second, he recommended that firm and drastic instructions of a menacing nature be personally delivered to Madero. Finally, when the State Department would not approve these severe recommendations, he personally led the members of the diplomatic corps in a joint venture urging Madero's resignation. At the same time, Wilson was in personal communication with General Huerta, leader of the federal forces in defense of Mexico City. On February 18, 1913, Madero was arrested through the treachery of General Huerta. Two days later, a provisional government was installed, with Huerta as president. In the early morning of February 23, Madero and his vice president were shot and killed while being taken under escort to the penitentiary from the National Palace. The following day, Ambassador Wilson telegraphed Secretary of State Knox that he was disposed to accept the Huerta government's version of the murders and to consider the incident closed. At the same time, he strongly recommended that the United States formally recognize the new dictator.[46]

Ambassador Wilson, with U.S. commercial interests at heart, had actively connived in the coups that deposed Madero. With Madero's death, the apostle of Mexican democracy had been destroyed. But his martyrdom would pass on to the succeeding phases of the revolution, and with it, the perception that foreign oil interests had participated actively in efforts to thwart the revolution.

Huerta's Dictatorship

Despite Wilson's urging, the United States never recognized the government of General Huerta. Led by the press, U.S. public opinion condemned the murder of Madero. Woodrow Wilson, whose inauguration to the U.S. presidency coincided with Huerta's ascendancy in Mexico, made his position clear in a pronouncement on March 11, 1913: "We can have no sympathy with those who seek to seize the power of government to advance their own personal interests or ambitions."[47]

This proclamation, which represented a significant change in U.S. policy toward Mexico, raises the issue of how President Wilson could reconcile his public views with the activities of his ambassador in Mexico City. The president distrusted Wilson from the start. Even before his inauguration, Wilson was fully briefed on the ambassador's activities during the Huerta takeover and on his reaction to the murder of Madero. President Wilson tolerated the ambassador for three months, but by June 1913, based on reports received from William Bayard Hale, his special emissary to Mexico, Wilson started to campaign for the ambassador's removal. In June, the president received a memorandum from Hale, which included the information that Ambassador Wilson had entertained General Huerta at the embassy for dinner. The president was shocked and passed the memorandum to Secretary of State William Jennings Bryan, with the annotation: "I think Wilson should be recalled."[48] On July 1, the president wrote a note to Secretary of State Bryan: "The document from Hale is indeed extraordinary. I should like . . . to discuss with you very seriously the necessity of recalling Henry Lane Wilson in one way or another."[49] Wilson wrote again to Bryan on July 3: "After reading Hale's report and the latest telegram from Henry Lane Wilson, I hope more than ever you will seriously consider the possibility of recalling Wilson."[50] On July 25, Wilson arrived in Washington, D.C., for consultation, and on August 4, Bryan informed him that the president would accept his resignation.

But there was a deeper issue here, which influenced the conduct of

U.S. diplomacy in Mexico and elsewhere throughout Wilson's administration. Both Wilson and Bryan came into office with progressive ideas, and they distrusted the silent, slow-moving, often extremely useful and powerful permanent bureaucracy of the State Department, which had been under Republican control for three administrations. On the part of the State Department, it was probable that the undersecretaries and experts looked with long-suffering indulgence bordering on indignation at President Wilson's pronouncements on Mexico, considering them to be impractical, amateurish, and idealistic. They would rather have played safe by following traditional policy. The president did not appreciate the need to reckon with the inert power of this bureaucracy. As early as March 20, 1913, Assistant Secretary of State Huntington Wilson tendered his resignation in protest of the president's public disavowal of U.S. participation in the Chinese Six Power Loan Consortium. It was a reversal of "dollar diplomacy," which had dominated the Taft and Roosevelt administrations, and no State Department diplomats had been consulted. Wilson was irate in his letter of resignation: "I had no reason to suppose that the officials on duty in the Department of State would learn first from the newspapers of a declaration of policy which I think shows on its face the inadequacy of the consideration given to the facts and theories involved."[51] It was partly for this reason that Wilson began to use extradiplomatic agents—secret ambassadors, such as Colonel Edward House, William Bayard Hale, and John Lind, all of whom Wilson trusted and who were to have a profound impact on Mexican affairs. Thus, the moral weight of Wilsonian policy was mobilized against the Huerta regime.

During this phase of the Mexican Revolution, foreign oil interests concentrated their attention in two areas: on the defense of their properties against depredation by federal and Constitutionalist (that is, rebel) forces, and on efforts to influence the policy of their home governments toward recognition of the Huerta regime. The theme of Anglo-American competition continued to dominate the Mexican oil market and heavily influenced the diplomacy of the period.

Within one month after Huerta assumed power, Carranza, governor of the state of Coahuila, commenced an insurgency on behalf of the Constitutionalist movement and soon was joined in rebellion by generals Villa and Obregón in the north and Zapata in the central states. One major objective for both federal and Constitutionalist forces was to obtain control of the rich area on the east coast from Tampico to Veracruz, which included all of Mexico's major oil fields. Control of these oil fields meant financial strength through the levy of special taxes and forces loans.

Doheny described his experience under the Huerta regime before the U.S. Senate's Foreign Relations Committee:

On May 15, 1913, Constitutionalist General Larraga appeared in Ebano . . . at the camp of the Mexican Petroleum Co. . . . with a force of 200 troops. He arrested the superintendent, took such supplies as he needed, made a forced loan of $5,000, and went away with all the rifles in camp, which rifles the company had secured for "protection" at the request of Hon. Ernesto Madero, Minister of Finance, under President Madero.

In October, 1913, the Huerta Government, through a packed and spurious supreme court, imposed a fine of $400,000 United States currency on the company, and threatened stoppage of operations in case of nonpayment. The company, through its representatives in Mexico, having in mind the policy of financial blockade then followed by the American Government, referred the question of paying this fine to the Hon. John Lind, personal representative of the President of the United States. On Mr. Lind's request to resist payment of this imposed "fine", [the company] did resist, at the risk of the destruction of its business and at the jeopardy of the liberty of its officials in Mexico, and succeeded in delaying settlement, which was still pending when Huerta was forced to leave Mexico.

In December, 1913, Constitutionalist Gen. Candido Aguilar appeared at the company's producing camp at Casiano . . . with a large armed force, demanding a loan of $10,000. He took supplies and all the rifles the company ever owned. . . . At the same time another of his bands appeared at the Potrero camp of the Eagle Oil Co. and demanded the same sum.

[Pearson's] Eagle Oil Company refused to make payment. Aguilar's men stopped the company's pumps, causing the oil and gas to break out around the well and under the Buenavista River. The escaping oil and gas have since been ignited by lightning and burned for three months. The well is forever in a dangerous condition by reason of the stopping of the pumps. . . .

[Our company] desirous of cooperating with the American Government in its Mexican policy, referred the Aguilar request to Mr. Lind, then in Tampico, through the American consul. Mr. Lind advised the company to pay the "loan", which it did promptly. Its pumps were not then and never after stopped.[52]

At this point in the testimony, Doheny interjected

At that time it was a well-known fact that the British assisted in the sale of a large amount of Huerta bonds and they were distinctly favorable to the Huerta Government at that time. Our Government had shown its animosity to Huerta and its desire to support his opponents. So that our action was in line with our own Government and that of the British was in line with the supposed sympathies of the British Government.[53]

This somewhat sanctimonious explanation of payments to rebel forces, while factually true, glossed over Doheny's real motivation—to continue pumping oil from his rich wells without allowing civil strife to interrupt its flow.

During the entire period of the Mexican Revolution, U.S. diplomacy was characterized by inconsistencies. Thus, while Wilson's special envoy, John Lind, was recommending payment of the rebel's demands, Admiral Frank F. Fletcher, in charge of U.S. naval forces off the Mexican coast, was warning the rebels against such forced levies. In any event, forced payments to rebel leaders, while particularly prevalent during the Huerta era, would continue under the Carranza regime. They became a pro forma method of operation for the oil companies and were seized upon by critics as proof of oil company support for antigovernment insurgencies. While there is no question that the monies that were paid assisted the rebellion, there is serious question as to the voluntary nature of the contributions.

Meanwhile, in the United States, businesses with big Mexican investments were exerting pressure on President Wilson to recognize the Huerta regime. When the futility of this approach became obvious, they campaigned for U.S. military intervention to protect their holdings. The situation was complicated even further on May 3, 1913, when Britain recognized General Huerta's provisional presidency. Crosscurrents of Anglo-American commercial rivalry in Mexico became so intermixed with the general diplomatic stream of affairs that the two became inseparable.

On May 12, 1913, a communication from Judge D. J. Haff of Kansas City on behalf of the chairman of the board of directors of the Southern Pacific Railway was delivered to President Wilson. Among other business leaders approving the letter was Doheny. The letter outlined the prevailing philosophy of the U.S. business community and presented certain proposals for maintaining influence in Mexico:

> The Constitutionalists steadfastly refuse to recognize Huerta or to treat with him. The United States government, therefore, has a great opportunity, by acting quickly, of presenting a plan to Huerta agreeing to recognize him on condition that he call an election at an early date. . . . We do not think it necessary to insist that Huerta shall resign and some other interim President be appointed in his stead. . . . He is the *de facto* President at the present time, and is a man of energy and executive ability, is in command of the army and is, better than any other person, able to carry out such an agreement. . . . In addition to that fact, foreign nations are becoming restive and are seeking to undermine the influence of the United States in Mexico. The British government has already recognized¨Huerta in a most marked manner by autographed letter from the King due to the efforts of [Pearson] . . . who has the largest interests outside of American interests in the Mexican republic. He is

using his efforts to obtain a large loan in England, and we are informed that he has succeeded on condition that the English government would recognize Huerta, which has been done. If Mexico is helped out of her troubles by British and German influence, the American prestige will be destroyed in that country and Americans and the commerce of the United States will suffer untold loss and damage.[54]

It is reported that President Wilson was sufficiently impressed by this letter's arguments to consider seriously conditional recognition of Huerta—the conditions being that all hostilities cease and that there be an early election. Ultimately, however, in view of Huerta's increasingly arrogant attitude toward the United States and his dictatorial methods, President Wilson decided against such a course.[55]

President Wilson believed that "big business" was complicating the diplomatic issue, as well as the entire course of events in Mexico. Josephus Daniels, secretary of the navy, reported in his diary: "The general opinion in the Cabinet was that the chief cause of this whole situation in Mexico was a contest between English and American Oil Companies to see which would control; that these people were ready to foment trouble, and it was largely due to the English Company that England was willing to recognize Mexico before we did."[56]

During this same period, Sir William Tyrrell, one of the most influential members of the British Foreign Office, visited the United States and had an interview with Secretary of State Bryan. In the course of their discussions on Mexico, Bryan, having already chastised Tyrrell concerning British imperialism,

proceeded to denounce Great Britain in still more unmeasured terms. The British, he declared, has only one interest in Mexico, and that was oil. The Foreign Office had simply handed its Mexican policy over to the "oil barons" for predatory purposes. "That's just what the Standard Oil people told me in New York," the British diplomat replied. "Mr. Secretary, you are talking just like a Standard Oil man. The ideas that you hold are the ones which the Standard Oil is disseminating. You are pursuing the policy which they have decided on. Without knowing it you are promoting the interest of Standard Oil.[57]

By July 1913, lobbying for various policies in Mexico had intensified to the point where President Wilson remarked to his private secretary: "I have to pause and remind myself that I am President of the United States and not of a small group of Americans with vested interested in Mexico."[58]

Despite this pressure, Wilson pursued a course of nonrecognition and pushed to unseat Huerta. As the conflict in the Mexican oil fields grew, the State Department requested that U.S. citizens leave the area for their own safety. But producing oil wells could not simply be abandoned, so the U.S. oil interests turned from lobbying for recognition of Huerta to urging that the United States provide military protection for their Mexican property and employees. The petroleum colony, strung along the Panuco River above Tampico, feared that whether on purpose or by accident, these strategic installations might be set afire in an engagement between federal forces constitutionalists. According to historian Howard Cline: "British, French, German and even Spanish warships converged to protect the threatened interests of their nationals. . . . Various national naval commanders warned both Constitutionalists and Federalists away from the oil installations. None had qualms about enforcing their warnings with action."[59]

On April 7, 1914, the United States received a telegram from its Consul at Tampico, stating that

> Waters-Pierce Refinery which had been occupied by attacking forces for last two days and as consequence has been under fire Federal gunboats and land forces with loss of oil tank by bursting of shells, fears loss to all refinery and demands protection.
>
> Warehouses of Agencia Comercial, German, located just below refinery were burned today. Loss one half million dollars. Situation as to foreign property complicated and serious.[60]

Secretary Bryan replied on April 9:

> In view of advices that the attacking force [Constitutionalist] has been occupying the Waters-Pierce Refinery, an American industry, and as a result has been drawing fire from Federal gunboats and land forces . . . jeopardizing the entire property of this and other foreign holdings . . . this Department will request the Secretary of the Navy to instruct the Admiral in command of the Gulf Fleet to confer with you to the end that the leaders of contending forces . . . should be notified and duly impressed with the fact that any destruction of the properties of these great foreign interests will produce a situation which will be a matter of grave concern to the United States and foreign governments and create a condition raising the question as to what measures this Government should take to secure such protection of the properties described as circumstances may require.[61]

The scene was set for military intervention. On the same day that Secretary Bryan sent his reply to Consul Miller, seven U.S. sailors were arrested in Tampico and subsequently released with verbal apologies. Admiral Henry T. Mayo demanded a formal apology including the firing of a 21 gun salute to the American flag which Huerta refused to do. Events culminated 12 days later when, ostensibly to prevent delivery of a shipment of arms to federal forces, the U.S. Marines landed at Veracruz.

They were opposed by a group of citizens and Mexican naval cadets who were subdued with 200 casualties. U.S. marines occupied Veracruz for the next seven months. Although all Mexicans united in decrying the U.S. seizure of Veracruz, Huerta was unable to direct this indignation into support. The incident triggered a mediation offer by the governments of Argentina, Brazil, and Chile, which Wilson converted into a hemispheric call for unseating the dictator. Huerta departed from Mexico in August 1914, leaving its government to Carranza, Obregón, Villa, Zapata, and the Constitutionalists.

Carranza and the Constitution of 1917

During the Huerta regime, the oil companies had struggled to defend their properties from the ravages of a civil war, and the U.S. government had provided protection, leading to military intervention and the dictator's eventual overthrow. Under Carranza the scenario differed. U.S. military intervention in Mexico occurred in the form of General Pershing's punitive expedition to find and punish Pancho Villa for his rampages across the Texas border. But this incident was not connected with oil. Instead, the oil interests conducted their struggle with the Carranza regime in Washington, D.C., London, and in the Mexican court system— against a backdrop of a world war, which was creating a new strategic demand for Mexican oil. The struggle consisted of sustained joint attacks by the oil companies against the Constitution of 1917, which had reinterpreted Mexican mineral law, striking at the ownership rights of the foreign concessionaires. In all of this, the U.S. government strongly supported the U.S. companies.[62]

Carranza's reinterpretation of Mexican mineral law was framed in the revolutionary rhetoric of the Constitution of 1917. It reinstated the essential elements of the traditional mining code of Spanish and Mexican law, which had governed the disposition of subsurface resources for centuries prior to the Díaz regime. Charles III had codified mineral law for *Nueva-España* in 1783:

> Mines are the property of my royal crown, both by their nature and their origin. . . . without separating them from my royal patrimony I grant them to my subjects . . . upon two conditions, that they shall contribute to my Royal Treasury the prescribed portion of metals; and second, that they shall work . . . the mines complying with what is prescribed. . . . whenever a failure shall occur in complying with those ordenanzas . . . they may be granted to any person who for that cause may denounce them.[63]

"Bitumens or juices of the earth" were included in the definition of minerals to which this law applied. In essence, traditional law declared that the title to mineral substances, including petroleum, was in the sovereign, who granted rights and governed the working of mines as he desired. Mexico's President Benito Juárez had reaffirmed the traditional law in 1863.[64] However, by 1884, with the encouragement of President Díaz, U.S. and British mining companies were investing heavily in Mexican mineral property, and the law required modification for their protection. First, Díaz's government amended the constitution to take mining matters into national jurisdiction. Next, a new mining code was passed into law:

> Article 6. Foreigners may acquire mining property on such terms and with such limitations as the law of the republic grant them the capacity to acquire, own and transfer ordinary property. . . .
> Article 10. The following substances are the exclusive property of the owner of the land, who may therefore develop and enjoy them, without the formality of entry or special adjudication.[65]

This mining code was further reinforced by the Mining Law of June 4, 1892:

> Article 4. The owner of the land may freely work, without a special franchise in any case whatsoever, the following mineral substances: mineral fuels, oils and mineral waters.
> Article 5. All mining property legally acquired and such as hereafter may be acquired in pursuance of this law shall be irrevocable and perpetual, so long as the Federal property tax be paid, in pursuance of the provisions of the law creating said tax.[66]

Finally, to avoid any ambiguity, the Mining Law of November 25, 1909, stated that "deposits of mineral fuels, of whatever form or variety" were the exclusive property of the owner of the soil.

With these laws clearly defining and defending the rights of private ownership, several oil companies were eager to purchase or lease Mexican land for oil exploration. Thus, the foreign oil interests were dismayed when Carranza's new constitution revoked all of the Díaz legislation, putting their title to Mexico's oil resources in serious jeopardy. Article 27 declared that the nation was the original owner of all lands and vested in the nation direct dominion over all minerals, including "petroleum and all hydrocarbons, solid, liquid or gaseous." It described this dominion as inalienable and imprescriptible and warned that concessions would be granted by the federal government only on condition that "regular works be established for the exploitation of the resources in question." Furthermore, only Mexicans by birth or by naturalization would have the right to obtain concessions to develop mines or mineral fuels in Mexico, and foreign corporations were specifically excluded from obtaining concessions under any circumstances. It stated that "the Nation may grant the same right to aliens, provided they agree . . . to be considered Mexicans in respect to such property, and accordingly not to invoke the protection of their Governments in respect to the same, under penalty, in case of breach, of forfeiture to the Nation of the property so acquired." [67]

Article 27 of the Constitution of 1917 should not have come as a surprise to foreign oil interests; its substance had been forecast many times in revolutionary speeches and in proclamations throughout the regimes of Madero, Huerta, and Carranza. From the Mexican perspective, Article 27 fulfilled two phases of the revolution: to reduce foreign ownership and enterprise in Mexican life, and to extend the supervision of the state, as an instrument of a popular revolutionary uprising, over the distribution and use of the mineral wealth of the country. [68] But to make its constitutional provisions effective, further legislation was required to define details and provide an administrative procedure. Such regulatory legislation would not be passed until 1925.

The U.S. oil interests were surprised by the Constitution of 1917. The fear of possible expropriation drove them once more to seek protection from the State Department and the White House. In reply to urgent inquiries, Ambassador Henry P. Fletcher was assured by Mexican authorities that Article 14 of the constitution offered safeguards against spoilation: "No law shall be given retroactive effect to the prejudice of any person whatsoever." [69] Based on this interpretation, as well as on further personal assurances that the ambassador obtained from Carranza in August 1917—that the Mexican government would not nationalize the oil industry and that U.S. properties, particularly oil holdings, would be protected—the Carranza regime was recognized in full and de jure by the United States. [70]

However, within a year, Carranza issued a series of executive decrees that appeared to justify the oil companies' worst fears. These included requirements that the government grant drilling permits for new wells (many of which were refused) and that petroleum land be registered with federal authorities. New taxes were imposed on oil lands held by leases made prior to May 1, 1917. These included an annual tax of from 5 to 50 percent of the rental and 5 percent of the royalties. Inherent in payment of rentals and royalties to the Mexican government was an acknowledgment that Article 27 was valid, that is, that the nation did in fact hold exclusive rights over subsurface resources.

The U.S. oil companies, claiming that the Carranza decrees were hampering their wartime output, sought military intervention at Tampico. They argued that if Carranza continued his course and nationalization occurred, oil would be cut off from the allies, since Mexico, as a neutral, would be required to embargo oil exports as a contraband of war.

Although never stated openly, this theme was behind a series of strong State Department notes delivered to the Mexican government during this period.*

Matters came to a head in the summer of 1918 at a White House conference. The secretary of the navy, Josephus Daniels, presented the oil company case for intervention.† The federal fuel administrator, John R. Garfield, and Bernard Baruch, of the War Industries Board, took an opposing view, arguing that oil producers were taking advantage of the war to get unwarranted concessions from Mexico. President Wilson resolved the controversy in a firm manner: "What you are asking me to do is exactly what we protested against when committed by Germany. You say this oil in Mexico is necessary for us. That is what the Germans said when they invaded Belgium—'It was necessary' to get to France. Gentlemen, you will have to fight the war with what oil you have."[71]

With Wilson's decision precluding military intervention to protect their property, the oil companies turned to the Mexican courts. During 1918-19 they filed more than 150 requests for injunctions halting application of the Carranza decrees. Judicial appeals became the modus operandi of the foreign oil companies in their struggle against the Mexican government during the later years of the war.[72]

The role played by Daniels in the intervention debate emphasizes the second major aspect in this phase of the oil struggle in Mexico—a rapidly escalating worldwide demand for petroleum products. As navy secretary, Daniels needed oil for ship propulsion. The Carranza regime not only coincided with World War I but with the conversion of the U.S. and Royal navies from coal to oil. In view of the wartime restriction of access to oil in the Middle East and Russia, the increased demand incurred by naval requirements could be met only by an increased supply of Mexican oil.

In Britain, Winston Churchill was First Lord of the Admiralty when the final decision to convert the Royal Navy from coal to oil was made.

---

*The State Department argued that the taxes and royalties were so burdensome as to be confiscatory. President Carranza defended them as necessary fiscal measures.

†Howard Cline is conclusive in declaring that Daniels was the spokesman for the oil interests at the White House meeting. Bernard Baruch, in his memoirs, is inconclusive, merely mentioning that one official proposed U.S. seizure of the oil fields at Tampico with squadrons of marines, who were already alerted. Daniels refers to the meeting in his diaries with an interesting but enigmatic denial complete with a question-mark: "Garfield & Requa seemed to lean toward the oil men, Baruch & I (?) not." In view of Daniel's cabinet responsibility for the navy during the conversion of the battleships from coal to oil, it would be logical that he would be seeking to protect sources of Mexican oil for naval use. (See Howard F. Cline, *The United States and Mexico* [Cambridge, Mass.: Harvard University Press, 1953], p. 187; Bernard M. Baruch, *Baruch, The Public Years* [New York: Holt, Rinehart & Winston, 1960], p. 83; and David Cronon, *The Cabinet Diaries of Josephus Daniels, 1913-1921* [Lincoln: University of Nebraska Press, 1963], p. 328.)

The Fast Battleship Division, under construction in 1912, required a relatively high speed in order to maneuver around the German battle fleet, and "we could not get the required power to drive these ships at 25 knots except by the use of oil fuel."[73] In making this decision, Churchill admitted that the Admiralty would have to contend with a sea of troubles. Changing the foundation of the Royal Navy from British coal to foreign oil, which was "in the hands of vast oil trusts under foreign control," posed almost insuperable problems in terms of guaranteeing oil in adequate supply from a secure source at an affordable cost.[74]

Seizing his opportunity, Pearson charged to the rescue. In June 1912, Pearson sent Churchill a letter, seeking an early interview, and stating:

> Privately, I may say that we are being pressed to sell the control of the Mexican Eagle Oil Company, Limited to one of the big existing Oil Companies. Should we do so the fuel oil supplies suitable for Admiralty purposes would thereby become controlled by a foreign Company, and it is this position that I should like to discuss with you.
>
> The Mexican Eagle Oil Company unquestionably owns the finest deposits of crude oil suitable for fuel purposes that today are controlled by any one Company.[75]

The oil company alluded to was Standard Oil of New Jersey, which was making another bid to purchase majority control of the Mexican Eagle Company. The interview eventually led to a substantial contract with the Admiralty for the purchase of Mexican oil. But the contract had to receive parliamentary approval as part of the naval estimates, and Pearson was not only an oil entrepreneur but a back-bencher of the then-ruling Liberal Party. Furthermore, other prominent members of the Liberal Party were stockholders in Mexican Eagle. This apparent conflict of interest caused a long and bitter debate in the House of Commons—but one that Churchill defused with his usual brilliant rhetoric. Pointing out that the price of oil (which in 1911–12 could compete on favorable terms with coal) had almost doubled in 1913 and that freights had risen by 60 to 70 percent, Churchill argued that it was necessary for the Admiralty to become an independent owner and supplier of petroleum.* However,

---

* The Admiralty ultimately did secure its own petroleum. Churchill established a Royal Commission on Oil Supply under Lord Fisher. After an on-site examination of the Persian oil fields by members of the Royal Commission, and with the cooperation of the governor of the Bank of England and the director of the Anglo-Persian Oil Company, Churchill negotiated through Parliament an Anglo-Persian Oil Agreement and Contract. Under the contract, for £5 million, His Majesty's Government obtained controlling shares in the Anglo-Persian Oil Company, which guaranteed the Royal Navy a substantial proportion of its oil supply at considerable economy. Ten years later, in 1923, Churchill estimated that the return on the government's initial investment in the Persian oil fields approached £40 million. (See Winston S. Churchill, *The World Crisis, 1911–1914* [New York: Charles Scribner's Sons, 1930], pp. 139–40.)

until that could be done, interim contracts for oil supply were necessary. Since Mexican Eagle was the largest British-owned oil company, it was an obvious candidate for such a contract. Addressing the conflict of interest issue leveled against Lord Murray, prominent Liberal Party member, Churchill concluded:

> But even if there were twenty Lord Murrays, and if every one of them had 20,000 shares, and if all the funds of the Liberal party, past present, and prospective, were exclusively invested in this company, we cannot see in what way these facts would be relevant to the decision which the Admiralty have to take, or how they could be held to debar us from doing what is profitable to the public and necessary for the Navy.[76]

With the approval of a major contract to supply Mexican oil to the Royal Navy, Pearson had the solid backing of the British Foreign Office in defending his Mexican oil properties against the decrees of Carranza.

Experiments in oil propulsion had been conducted in the United States since 1900, and most smaller ships had already been converted. However, the decision to convert battleships to oil propulsion was made only after successful tests on the battleship *Idaho* in 1912. Nevertheless, because there was no real need, President Wilson eventually decided not to pursue Mexican oil through military intervention. Unlike Great Britain, the United States had an indigenous supply of oil, and the conversion of its navy for coal to oil did not entail its trading abundant domestic fuel for a scarce one. Moreover, in spite of dire predictions, Mexican oil remained available. The Carranza regime, bowing to the protests of foreign governments, did not enforce nationalization. Though plagued by red tape, constant doubt, and continuous court battles, new wells were drilled, and there was a constantly increasing flow of oil from the rich Mexican fields.

In the midst of revolution and reform, the golden age of Mexican oil occurred between 1917 and 1921. While other foreign industries in Mexico languished or departed in the chaos of revolutionary activity, oil prospered. Mexican oil production soared from 500,000 barrels in 1906 to 55 million barrels in 1917.[77] Between 1918 and 1921, oil production of Pearson's Mexican Eagle Company doubled, from 16.8 million to 32.4 million barrels.[78] Pearson sold the Mexican Eagle Company in 1919, at the height of its prosperity, to Royal Dutch Shell.

Why did oil succeed in Mexico despite the revolution? Part of the reason had to do with the sheer wealth of the resource and the fanatical perserverence on the part of a very special group of entrepreneurs. A more important reason was the strategic interests of allied governments during World War I. Mexico produced nearly 25 percent of the world's oil supply and was vital for fueling the allied war effort. There is little doubt that the strong representations of the U.S. State Department and the

British Foreign Office influenced Carranza to delay his confiscatory legislation. Official home government support and protection appears to be the most important reason for the continued success of the foreign oil enterprises in Mexico. No matter what the situation, either by diplomatic note, threat, or by actual intervention, the United States and Great Britain prevailed against the revolutionary governments of Mexico in defense of the rights of foreign oil interests on Mexican soil.

Official intervention by two major foreign powers in the internal affairs of Mexico combined with the questionable actions and ethics of foreign oil companies and individual entrepreneurs forms the foundation for a still-active distrust of foreign oil interests by Mexicans and for continued opposition to their participation in the Mexican oil economy. It is an active form of economic nationalism that would sacrifice substantially higher profits in order to prevent foreign penetration of the new oil market.

Between 1910 and 1920, a series of events helped to subvert Mexico's revolution and to fuel a negative perception of foreign oil interests: Standard Oil's financing of the insurgency against Díaz; Pearson's role in Madero's downfall; the U.S. ambassador urging a Mexican president to resign; the U.S. Marines occupying Veracruz; and U.S. oil companies making continuous payments to insurgents of every stripe. Whether true or not, the perception persists that foreign oil interests created and helped to destroy the regimes of Madero and Huerta. Today, when the revolution is enshrined and Madero and Carranza are considered the apostles of their country's democracy, Mexico continues to resist the basic counter-revolutionary bias of foreign enterprise.

The discovery and exploitation of Mexican petroleum resources by foreign interests is only the first part in the background of Mexican oil. The second part, dealing with the manner in which the reform mandates of the revolution were eventually implemented and chronicling the events leading to the expropriation of foreign oil interests in 1938, is equally important for understanding economic nationalism in Mexico today.

THE EXPROPRIATION OF 1938

The Obregón-Calles Years: 1920–35

The last successful military uprising in Mexican history took place in 1920 when Venustiano Carranza was overthrown by the Agua Prieta, led by Álvaro Obregón and Plutarco Calles. The "Sonora dynasty," which they created, was to have effective control over Mexico until 1935, when Calles was run out of the country by Lázaro Cárdenas.

After a brief interim presidency (by Adolfo de la Heurta), General Obregón began a four-year term in 1920. He was succeeded in 1924 by Calles. Mexico's first postrevolutionary leaders, particularly President Calles, inherited not only the fervor and rhetoric of a socialist-oriented revolution but the harsh economic realities, which made the ambitious goals of the revolution seem nearly unattainable. Calles, well aware of national feeling about placing the control of natural resources in the hands of foreigners, played a highly cynical but largely successful game for more than a decade: he criticized the foreign oil companies while simultaneously creating a legal basis for their continued operation in Mexico. In this way, he was able to encourage the economic development of Mexico's oil regions—and earn some extra income for the highways, schools, and other welfare programs sanctioned in the revolutionary constitution—while assuaging the vociferous left wing in Mexican politics, which demanded the removal of foreign oil companies.

Until the late 1920s, the dual policy approach of Calles diverted Mexico's attention away from the crucial question of subsoil ownership of oil reserves. However, as political corruption and the failure of land distribution policies became clearer, criticism of foreign oil operators again burgeoned. Calles, always a political realist, promulgated the so-called Calles-Morrow Agreement of 1927, which guided the relations between the oil companies and the Mexican government until the late 1930s.

The Calles-Morrow Agreement was a stopgap measure. The U.S. State Department had long defended the oil companies against the application of Article 27, which reserved subsoil resources for the Mexican state, by claiming that concessions granted before the revolution could not be altered by ex post facto legislation.[79] President Calles had tried to circumvent this issue in 1925 by passing a law that required the oil companies to obtain new concessions, limited to 50 years duration, to replace their pre–Constitution of 1917 unlimited duration agreements.[80] The United States opposed this measure, arguing that the new law imposed such a fundamental change in the property rights of the oil companies that it constituted confiscation.[81] U.S. capital reacted by reducing Mexican investments, and economic contraction ensued. Thus, it was to protect his increasingly conservative regime that Calles met with the U.S. ambassador to Mexico, Dwight Morrow, and concluded the Calles-Morrow Agreement, modifying the status of the oil concessions.[82] The Petroleum Act of 1925 was amended to allow "confirmatory concessions" of unlimited duration to be issued upon application to those owners and leaseholders who had acquired their subsoil concessions prior to May 1917.[83]

Calles became more conservative after reaching the agreement with Morrow. Lorenzo Meyer has written:

Starting in 1928, right after reaching the agreement with Morrow, Calles seemed to change his outlook in a number of ways: he broke openly with the labor movement, lost all interest in the agrarian reform, and frankly devoted himself to developing the country in the traditional way, with himself and his group as the main promoters. . . . Whereas during the first years of his government Calles had redistributed more land than any of his predecessors, at the end he tried to put a total stop to the agrarian reform.[84]

In 1928, it was General Obregón's turn to reassume the Mexican presidency. Obregón easily won the election, but he was murdered by a religious militant just prior to taking office. No longer in office, Calles formed the Partido Nacional Revolucionario (PNR) in 1929 to bring together all factions of the governing group. In his role as the PNR's chief and "Supreme Leader of the Revolution," Calles was able to dominate Mexican politics until 1935.[85]

The PNR appeared to be a highly democratic, if centralized, organization, with local caucuses and representatives. Although it conveyed a surface appearance of progressiveness, in practice, the PNR enabled Calles to weed out opposition and consolidate his rule. In cases where anti-Calles candidates for local office managed to squeeze through the party net, polling places were attended by military squads sporting machine guns.[86] As one U.S. writer observed:

The Mexican Revolution, 1932: Democracy, that had never been, was not. Land distribution, that had amounted to very little, had come to a virtual standstill. Labor reform reached a standstill and the government sided with employers in a controversy over state laws permitting expropriation of factories that had violated labor regulations. If wages had gone up, living costs had gone higher. The real condition of the peasants and the workers had not improved.[87]

The new PNR failed to assuage some generals. Before long, left-wing agitation reappeared in Mexico, and there was renewed talk of military rebellion. Behind the scenes, there was a desperate search to find a candidate who could return to the aspirations of the revolution and still preserve some of the stability of the Calles years. Lázaro Cárdenas became the chosen successor.

Lázaro Cárdenas

Though he had kept a low profile during the 1920s, Cárdenas came to be identified as a staunch advocate of returning communal lands to the Indian villages; he had defended this policy even when reactionaries under Calles had declared it a failure. Cárdenas was also a stubborn advocate of the rights of labor. In 1932, as governor of the state of Michoacán, Cárdenas had actually defied Calles by passing a law that permitted the expropriation of factories which had shut down rather than conform to labor regulations. President Calles, along with many other politicians trapped in the increasingly unpopular webbing of the PNR, began to think that Cárdenas might be the man to give a legitimate "face-lift" to their political machine. Mistakenly believing he could manipulate Cárdenas to preserve the regime he had built, Calles endorsed his presidential candidacy.

Cárdenas was especially sensitive to the decline in Mexican idealism throughout the last years of Calles's domination. He believed that his country needed a restorative dose of Mexican nationalism. Thus, it would be ingenuous to view Cárdenas's nationalization of foreign oil companies in 1938 as an unpremeditated action.

It is against this backdrop of Cárdenas's reforming spirit and the political and social milieu in post-Calles Mexico that the oil nationalization must be understood. The incessant legal maneuverings that were to dominate discussion of the expropriation crisis in the United States continually missed the point.[88] Mexico's move to gain control of its own natural resources was part of its ambition to fulfill its revolutionary goals. The revolution had been fought for a variety of complex and often contradictory reasons, but one thing is certain: Porfirio Díaz's granting of enormous concessions to foreign investors crystallized the concept of "Mexico for the Mexicans" in the Constitution of 1917.

As president, Cárdenas encouraged Mexico's developing nationalism based on the Constitution of 1917. For the first time since the Calles-Morrow Agreement in 1927, the petroleum industry had reason to worry. Cárdenas was not an easily corrupted bureaucrat. His revolutionary fervor, augmented by his immense personal popularity, sustained the coordinated effort by Mexicans of all political hues to regain control of the most basic aspects of the Mexican economy. The Constitution of 1917, guaranteeing both the right of labor to organize and bargain with employers and the ownership of subsoil resources for the state, was the foundation upon which Cárdenas sought to reconstruct Mexico's revolutionary ideals.

In 1936, the Cárdenas government embarked on a series of steps that directly challenged the position of the foreign oil companies. The first step was to propose wider distribution of national wealth by giving the Mexican government the right to expropriate any property whatsoever for "public use," with compensation to be paid in accordance with its "fiscal value," as determined by an appropriate court.[89] Though opposed by nearly all Mexican industry, the legislation passed.[90] The United States sent its ambassador, Josephus Daniels, to confront President Cárdenas directly about its worries. Cárdenas sought to assuage concern by emphasizing that the law was only to be used "for the nation" (a phrase that was deliberately left vague) or when dealing with companies or individuals who performed "basic" services" and who, for one reason or another, had decided to suspend operations. Nevertheless, it was clear that Cárdenas was beginning to act on revolutionary rhetoric in a way that had not been seen since the early days of the Obregón presidency.

At the same time Cárdenas was pushing his expropriation legislation through congress, Mexican labor was becoming a growing force. Article 123 of the Constitution of 1917 had guaranteed workers the right to organize, bargain, and strike. Yet, under Calles, the Mexican government had grown increasingly supportive of factory owners. Cárdenas's succession marked the first time since the mid-1920s that Mexico's workers received active support from the federal government. The new president made it clear he would support labor's efforts to redress grievances with foreign oil companies in the federal courts.

The emerging conflict had two principal themes: the legal definition of the rights of the state in the petroleum regions and the development of an oil workers' union able to bargain with the oil companies. The events of late 1936 determined that the unions would be the main threat which the oil companies would face to their continued operations in Mexico.

In early 1936, the fragmented oil unions united into one, Sindicato de Trabajadores Petroleros de la República (Union of the Oil Workers of the Republic), or STPRM. Relying on labor laws passed several decades earlier, STPRM presented a series of demands to the oil companies and a draft collective contract. The companies agreed in principle to sign a contract but balked at the 40-million peso wage hike the workers were seeking. No agreement was reached. On May 28, 1937, STPRM proclaimed an industry-wide strike. The confrontation so long feared by both the oil companies and the government had finally come. The strike of May 1937 was not simply a fight for better wages, benefits, and working conditions but was the first real attack on what Mexicans viewed as economic imperialism, with nothing less than the national honor at stake.

Almost a full year would elapse between the strike declaration and President Cárdenas's speech to the nation on March 18, 1938, which bluntly ended foreign participation in the Mexican oil industry. In the interim, the companies, the Mexican government, and labor organizations were locked in a dance of futile legal maneuver. Too much passion had been aroused; only an act as wrenching as expropriation would satisfy the mood of Mexico.

After the strike was called, the Mexican public, outraged by the companies' attempts to discredit the workers through full page newspaper announcements, began to lose patience quickly. The federal government ruled that the oil companies were obliged to pay the workers their regular wages during their absence from their jobs.[91] President Cárdenas made his position clear in a speech at Monterrey, warning all industrialists that if the dispute between labor and management persisted, either the workers or the government would take over. The management of the U.S. oil companies, organized in the Association of Producers of Petroleum in Mexico, informed the U.S. government that it was hoping for the best but feared the worst. Ambassador Daniels, a Roosevelt appointee sympathetic to Cárdenas, did not believe the government would nationalize the fields, but he did think the companies would have no choice but to grant wage increases.[92]

The strike could not last long. Mexico's economy was still dominated by struggling industrial enterprises and agriculture, and some immediate resolution of the oil crisis was necessary to prevent wholesale havoc. Since Mexico was exporting 60 percent of its oil production, the oil crisis threatened to lead to the quick exhaustion of the country's reserves of foreign exchange. Before the economy had a chance to be endangered, the striking unions petitioned the Federal Board of Conciliation and Arbitration to declare the dispute a national emergency.[93] The petition was accepted immediately, and the workers returned to work while the board prepared its report.

The board issued a 2,700-page report, which presented the government's view on all aspects of the present state of the Mexican oil industry. The report's 40 conclusions emphasized the discrepancy between Mexico's economic needs and the oil firms' policies and pointed to a large number of fiscal and political irregularities.[94] As far as the labor problem was concerned, the board's experts concluded that the companies were able to pay a wage increase of 26 million pesos a year—12 million more than they had been prepared to give.[95] The oil companies challenged these conclusions, countering that the real increases in wages would be 41 million pesos. At this point, they began to rally the foreign press and paint a dim picture of the justice of the Mexican action. President Cárdenas, for his part, encouraged the workers to press the companies for full compliance with the board's recommendations. After four months of review and testimony by both sides in the dispute, the oil companies, claiming they could not comply with the contract demands, appealed to the Mexican Supreme Court for an injunction.

The companies apparently anticipated that the Court's award might go against them. Already they had begun a financial offensive against the government in October 1937, trying by every means to deplete the monetary reserves of the Bank of Mexico.[96] Rumors spread that the peso, then being exchanged at the rate of 3.6 to the dollar, would soon be devalued. President Cárdenas, speaking before the STPRM congress on

February 24, 1938, charged that

> the recent attitude of the oil companies with respect to the dispute with
> their workers appears to indicate an effort, by the sudden withdrawal of
> their deposits and a tremendous press campaign, to foment alarm
> among businessmen and deny credit to industry, to make use of illicit
> coercion, in order to influence the character of the final decision on
> behalf of their commercial interests and to prevent a normal and upright
> conclusion of the case now before the legal authorities.[97]

On March 1, 1938, the Mexican Supreme Court upheld the Federal
Board of Conciliation and Arbitration's decision, reiterating what the
board had stated previously:

> The Commission asserted that the companies were able to increase their
> disbursements for wages and social services to workers and employees
> as demanded by labor. . . . The experts were unable to accept the data
> of the companies' books, and they impugned the accuracy of these
> accounts because they found that in many cases there had been
> concealment of profits by means of numerous subterfuges and book-
> keeping tricks, undoubtedly with the purpose of evading payment of
> federal taxes.[98]

The companies, hoping a less costly bargain could be struck, refused
to comply with the Supreme Court's ruling. This placed the Cárdenas
government in the position of having to enforce the Court's decision. The
companies, according to the Mexican viewpoint, had exhausted their
legal remedies and were in contempt of court when they refused to heed
the verdict. The oil workers, feeling they had no other recourse and
encouraged by the legislation concerning the right to expropriate factories, suspended work throughout the industry on March 18, 1938. That
evening, President Cárdenas signed the order expropriating the foreign
oil companies. Cárdenas had summarized his feelings in his private diary
a few days earlier:

> Today Mexico has its great opportunity to shake off the political and
> economic yoke that the oil companies have placed upon us while
> exploiting one of our most important resources for their own benefit and
> holding back the program of social reform set forth in the Constitution.
> Various administrations since the Revolution have attempted to do
> something about the subsoil concessions being enjoyed by foreign firms,
> but up until now domestic problems and international pressure have
> mitigated against this effort. Today, however, the circumstances are
> different: there are no internal struggles and a new world war is about to
> begin.[99]

The newly nationalized U.S. and British oil companies immediately sought support from their governments. They appealed to the Mexican courts, again, to no avail. They began to publicize the "fascist" or "Communist" government of Mexico, depending on their audience. And perhaps most important, they instigated a worldwide boycott of Mexican oil. However, President Cárdenas had judged the situation correctly: a more auspicious time could not have been selected to fulfill the revolutionary goal of a nationalized oil industry. War was beginning in Europe, and when Great Britain broke diplomatic relations with Mexico, Cárdenas began to flirt with German and Italian markets for Mexican oil. The United States, confronted with Cárdenas's opportunity to sell Mexican oil to the Axis, reacted less vehemently than it might have, and in an exchange of notes, it emphasized the common interest in hemispheric defense.[100]

The nationalization of 1938 is the single most important date in modern Mexico's petroleum history. The presidential decree of expropriation removed foreign participation from the Mexican oil industry at a stroke. All capital goods belonging to the companies were declared to be property of the nation. Since more than 90 percent of the industry's production capacity was affected by the decree, Mexico became the master of its own petroleum house overnight.

An economic interpretation of the events leading to the expropriation of nearly the entire oil industry will never fully explain what occurred on March 18. Something much more profound was at work, something born in the revolution, codified in the Constitution of 1917, but then laid aside during the ensuing years. A new mood, a new spirit of economic nationalism, was on the rise. It was this spirit that finally permitted Mexico to stand up and identify itself much in the way an adolescent chooses to separate from parental influence that has been perceived as undue. The attitudes formed as a result of the history of foreign participation in, and dominance of, the petroleum industry would remain in the Mexican mind for years. But for the moment, the new freedom from foreign influence made the risk of self-sufficiency sweet, and it was a joyous proud nation that awoke to March 19.

NOTES

1. Mira Wilkins, *The Emergence of Multinational Enterprise: American Business Abroad from the Colonial Era to 1914* (Cambridge, Mass.: Harvard University Press, 1970), p. 116.

2. Ibid.,

3. Ibid., pp. 118, 120.

4. Ibid., p. 116.

5. Ibid.

6. Peter Calvert, *The Mexican Revolution, 1910-1914* (Cambridge: Cambridge University Press, 1968), p. 20.

7. Ibid., p. 17.

8. Wilkins, op. cit., p. 119.

9. Pan American Petroleum and Transport Company, *Mexican Petroleum* (New York: PAP&TC, 1922), p. 23. Chapter 2, "History and Future of the Oil Industry in Mexico," is the text of an address delivered by Edward L. Doheny at the Second Annual Meeting of the American Petroleum Institute, held at the Congress Hotel, Chicago, December 6, 1921.

10. Ibid., p. 25.

11. Ralph W. Hidy and Muriel E. Hidy, *Pioneering in Big Business, 1882-1911* (New York: Harper & Brothers, 1955), pp. 128, 258, 464.

12. Arthur Pound and Samuel Taylor Moore, eds., *They Told Barron: Conversations and Revelations of an American Pepys in Wall Street* (New York: Harper & Brothers, 1930), p. 140.

13. Pan American Petroleum and Transport Company, op. cit., p. 17.

14. Ibid., p 26.

15. Ibid., p. 32.

16. Ibid., p. 35.

17. Ibid.

18. U.S., Congress, Senate, Committee on Foreign Relations, *Investigation of Mexican Affairs*, 1920, 66th Cong., 2d sess., document no. 285, pp. 218-19.

19. Alfred Tischendorf, *Great Britain and Mexico in the Era of Porfirio Díaz* (Durham, N.C.: Duke University Press, 1961), p. 67.

20. Edward I. Bell, *The Political Shame of Mexico* (New York: McBride, Nast, 1914), p. 126.

21. Tischendorf, op. cit., p. 125.

22. Bell, op. cit., pp. 126-27.

23. Charles P. Howland, *Survey of American Foreign Relations* (New Haven, Conn.: Yale University Press, 1931), p. 126.

24. J. Fred Rippy, *British Investments in Latin America, 1822-1949* (Minneapolis: University of Minnesota Press, 1959), p. 101.

25. Pan American Petroleum and Transport Company, op. cit., pp. 96-108.

26. Calvert, op. cit., p. 21.

27. Ibid., p. 18.

28. William Franklin Sands, *Our Jungle Diplomacy* (Chapel Hill: University of North Carolina Press, 1944), p. 146.

29. J. Fred Rippy, José Vasconcelos, and Guy Stevens, *American Policies Abroad, Mexico* (Chicago: University of Chicago Press, 1928), p. 103.

30. Wilkins, op. cit., p. 129.

31. Ibid., p. 278.

32. Ibid., fn. 129.

33. U.S., Congress, Senate, Hearings of Subcommittee of Committee on Foreign Relations, *Revolutions in Mexico*, 62d Cong., 2d sess., 1913, p. 548.

34. Calvert, op. cit., pp. 74-75.

35. Ibid., p. 76.

36. Wilkins, op. cit., p. 277.

37. Pound and Moore, op. cit., p. 141.

38. Lorenzo Meyer, *Mexico and the United States in the Oil Controversy, 1917-1942* (Austin: University of Texas Press, 1976), p. 27.

39. Rippy, Vasconcelos, and Stevens, op. cit., p. 107; Wilkins, op. cit., p. 131.

40. J. A. Spender, *Weetman Pearson, First Viscount Cowdray* (London: Cassell, 1930), pp. 189-90.

41. Calvert, op. cit., pp. 98–99.

42. Ibid., pp. 100–01.

43. Ibid., p. 109.

44. U.S. Department of State, *Foreign Relations of the United States, 1912*, (Washington: U.S. Government Printing Office, 1919), pp. 842–46.

45. Rippy, Vasconcelos, and Stevens, op. cit., p. 32.

46. Ibid., p. 37. For the text of various pieces of correspondence between Mexico City and Washington, D.C., documenting the ambassador's views, see U.S. Department of State *Foreign Relations of the U.S., 1913*, (Washington: U.S. Government Printing Office, 1919), pp. 731–36. See also Calvert, op. cit., p. 157.

47. Roy Stannard Baker, *Woodrow Wilson, Life and Letters* (Garden City, N.Y.: Doubleday, Doran, 1931), 4:71.

48. Ibid., p. 242.

49. Ibid., p. 255.

50. Ibid., pp. 257–58.

51. Ibid., p. 262.

52. U.S., Congress, Senate, *Investigation of Mexican Affairs*, pp. 283–84.

53. Ibid., p. 284.

54. Baker, op. cit., pp. 246–47.

55. Ibid., pp. 248–49.

56. David Cronon, *The Cabinet Diaries of Josephus Daniels, 1913–1921* (Lincoln: University of Nebraska Press, 1963), p. 43.

57. Burton J. Hendrick, *Life and Letters of Walter H. Page* (Garden City, N.Y.: Doubleday, Page, 1925), 1:203.

58. J. P. Tumulty, *Woodrow Wilson As I Knew Him* (Garden City, N.Y.: Doubleday, Page, 1921), p. 146.

59. Howard F. Cline, *The United States and Mexico* (Cambridge, Mass.: Harvard University Press, 1953), p. 155.

60. U.S. Department of State, *Foreign Relations of the United States, 1914*, (Washington: U.S. Government Printing Office, 1922), p. 668.

61. Ibid.

62. Howland, op. cit., p. 127.

63. Ibid., pp. 128–29.

64. Ibid., p. 130.

65. Association of Oil Producers in Mexico, *Documents Relating to the Attempt of the Government of Mexico to Confiscate Foreign-Owned Oil Properties* (February 1919), p. 2.

66. Ibid., p. 3.

67. Ibid., p. 4.

68. Howland, op. cit., p. 135.

69. Association of Oil Producers in Mexico, op. cit., p. 6.

70. Cline, op. cit., pp. 186–87. The Carranza government received formal recognition from the United States on August 31, 1917.

71. Bernard M. Baruch, *Baruch, The Public Years* (New York: Holt, Rinehart & Winston, 1960), p. 83.

72. Cline, op. cit., p. 187.

73. Winston S. Churchill, *The World Crisis, 1911–1914* (New York: Charles Scribner's Sons, 1930), p. 133.

74. Ibid., p. 138.

75. Randolph Spencer Churchill, ed., *Companion Volume II, Part 3* [this volume is a companion to Winston S. Churchill, *The World Crisis, 1911–1914*] (Boston: Houghton-Mifflin, 1969), p. 1930.

76. Great Britain, *Parliamentary Debates* (Commons), 5th ser., 60 (1913), 1479.

77. Howland, op. cit., p. 126.

78. Walter R. Skinner, ed. *Oil and Petroleum Manual, 1922* (London: Walter R. Skinner, 1922), p. 128.

79. Virginia Prewett, *Reportage on Mexico* (New York: E. P. Dutton, 1941), p. 71.

80. Ibid., p. 72.

81. Ibid.

82. Ibid.

83. Meyer, op. cit., pp. 132–38.

84. Ibid., p. 140.

85. Ibid.

86. Prewett, op. cit.,,p. 78.

87. Ibid.

88. See, particularly, Standard Oil of New Jersey, *Confiscation or Expropriation? Mexico's Seizure of the Foreign-Owned Oil Industry* (New York: SONJ, 1940); and Standard Oil of New Jersey, *Denials of Justice* (New York: SONJ, 1938–40).

89. Armando Maria y Campos, *Mugica: Cronica biografica* (Mexico City: Compañia Editiones Populares S.A., 1939), pp. 296–97.

90. Association of Producers of Petroleum in Mexico, *Documents with Regard to the Draft of the Expropriation Law* (Mexico City: APPM, 1936).

91. Antonio J. Bermúdez, *The Mexican National Petroleum Industry: A Case Study in Nationalization* (Palo Alto, Calif.: Institute of Hispanic American and Luso-Brazilian Studies, Stanford University, 1963), p. 12.

92. Meyer, op. cit., pp. 148–72.

93. Ibid.

94. Bermúdez, op. cit., p. 14.

95. Meyer, op. cit., p. 158.

96. Jesús Silva Herzog, *Petroleo Mexicano* (Mexico City: Fondo de Cultura Económica, 1941), pp. 18–19; Meyer, op. cit., p. 306.

97. Bermúdez, op. cit., p. 15.

98. Ibid.

99. Meyer, op. cit., pp. 167–68.

100. Ibid., p. 193.

3

THE PETROLEUM POTENTIAL OF THE REFORMA AND CAMPECHE OIL TRENDS

There no longer remains any doubt that southeastern Mexico's Reforma and Campeche trends (see Map 3.1) have enormous petroleum potential—so enormous that their speedy development could precipitate substantial shifts in the prevailing patterns and terms of the world oil trade. In order to better assess this potential from a policy perspective, it is useful to have at least a rough idea about the nature and magnitude of these resources as they relate to three considerations. The first consideration is the likely range of Reforma's and Campeche's commercially producible petroleum reserves and the speed at which new production can be developed. The second consideration is the probable division of these reserves between crude oil and natural gas. The third consideration is the probable range of the total per barrel costs of producing this petroleum.

Unfortunately, the discoveries are far too recent to allow a reliable, detailed determination of the extent and range of Mexico's petroleum reserves or to precisely estimate costs. It will be several years before Pemex has drilled the exploratory wells necessary to delimit the geographical and geological boundaries of the Reforma and Campeche trends and to assess the quantity of the petroleum reserves located in the several hundred oil-bearing structures already known to lie within them. Moreover, even when this firsthand information is established, access will be at Pemex's discretion.* This chapter, while recognizing the limitations

---

Michael Metz co-authored the section entitled Pemex's Data and Projections; Dean Goodermote co-authored the section entitled Estimates of the Average Total Cost of Producing Southeastern Mexico's Crude Oil.

*Pemex is the ultimate source of most of the production and cost data reported in the oil trade press. There have been allegations that some of these reports exaggerate the petroleum potential of southeastern Mexico. While this might be true in specific cases, the general trend of events does not appear to support this allegation.

Map 3.1: Reforma and Campeche Trends

GULF OF MEXICO

YUCATAN

CAMPECHE

BAY OF CAMPECHE

CAMPECHE

Ciudad del Carmen

TABASCO

Cardenas

Villahermosa

VERACRUZ

REFORMA

OAXACA

CHIAPAS

GUATEMALA

GULF OF TEHUANTEPEC

PACIFIC OCEAN

Source: Constructed by the author.

and reliability problems associated with using potentially self-serving official statements, data, and projections, nevertheless, will attempt to arrive at plausible rough estimates of the quantities and likely costs of Pemex's new petroleum reserves.

PEMEX'S DATA AND PROJECTIONS

Between 1961 and 1974, Pemex's estimates of Mexico's proved petroleum reserves (which always combine crude oil and natural gas) were remarkably stable—totaling 5 to 6 billion barrels. But within three years of the initial Reforma discovery, Mexico's proved reserves began to rise rapidly. Proved reserves—listed as 6.3 billion barrels as of year-end 1975—jumped to 11.2 billion barrels as of year-end 1976, to 16.8 billion barrels by year-end 1977, and to 20 billion barrels by September 1978.[1] Moreover, according to Pemex, Mexico's proved reserves will experience further rapid growth—to at least 30 billion barrels in 1982.[2]

Since Pemex takes the cautious position that only reserves from fields actually being produced can be recorded as proven, these estimates and projections of Mexico's proven reserves greatly understate its ultimate petroleum potential. To illustrate, the 1977 proved reserve estimates include 10 billion barrels in Mexico's southern zone (that is, Reforma and Campeche). But this takes account of reserves in only seven of the Reforma and Campeche structures—only one of which is thought to be fully developed.[3] No account had been taken of any petroleum in either the 18 other structures known (at that time) to be oil-bearing but not yet in production or in the several hundred additional structures that had been detected.[4] Given that Pemex claims to have maintained an 82 percent success rate during four years of exploratory drilling in the area, the probability of finding substantial new oil-bearing structures must be rated as high.[5]

Sharp expansions in Mexico's proved petroleum reserves have allowed a rapid expansion of her crude oil output. Beginning in 1974, Mexico's production of crude oil, which had been rising slowly but steadily since World War II, began to soar—from an average of 500,000 barrels per day in 1972, to 650,000 barrels per day in 1974, to 1.1 million barrels per day in 1977. Natural gas production has not enjoyed a comparable increase over this period—indeed, it has been stagnant. The reason is not a shortage of new reserves but rather a temporary paucity of markets (see chapter 7).

In December 1976, Pemex unveiled what at the time was labeled an ambitious $15.5 billion six-year program to accelerate petroleum exploration, development, and production. The plan called for more than doubling crude oil production to 2.2 million barrels per day in 1982 and nearly doubling natural gas production to 3.6 billion cubic feet per day. These plans were revised sharply upward in March 1978.[6] At the celebration of the fortieth anniversary of Mexico's nationalization of its oil industry, Pemex announced that its 1982 target for crude oil production would be reached by year-end 1980. Moreover, it projected that crude oil production would average 2.7 million barrels per day in 1982, and assuming the U.S. government allows the importation of substantial quantities of natural gas, Mexico's natural gas production would soar to 5.6 billion cubic feet per day. Based on southeastern Mexico's large petroleum reserves and Pemex's post-Reforma record of always exceeding its published production projections, these updated production targets seem plausible. If they are met, Mexico's daily exports of crude oil and natural gas will average about 1.5 million barrels and 2 billion cubic feet, respectively, in 1982.*

GEOLOGICAL AND GEOGRAPHICAL CHARACTERISTICS OF SOUTHEASTERN MEXICO'S OIL-BEARING STRUCTURES

Reforma

Reforma's oil fields have their origins in the Jurassic and Cretaceous geological ages and lie at a depth of 12,000 to 14,000 feet—far deeper than Mexico's previously discovered oil reserves. An unusual feature of the Reforma oil reserves is their high proportion of associated natural gas. Natural gas appears to account for nearly half of the petroleum potential in some of these fields.[7] Pemex's southern zone exploration manager has said:

> The best way to visualize a typical Reforma or Bay of Campeche field is to imagine a sealed and pressurized tank full of coarse gravel, with a bottom layer of water and then full of oil to the top. The vertical and horizontal flow channels mean the individual wells can sustain high production rates with little if any pressure drop when aided by water-flooding.[8]

Reforma's oil is contained in highly faulted sedimentary rocks, sandstone, and clays.[9] Its large reserves stem from the great volume and number of oil-bearing reservoirs: the thickness of the known oil-bearing

---

*Natural gas exports of 2 billion cubic feet per day are equivalent to about 300,000 barrels per day of crude oil.

strata ranges from 600 to 6,000 feet.* Moreover, Reforma's geographical boundaries are large and still expanding rapidly. Four strikes in early 1977 extended the producing trend 19 miles to the south, and a 19-mile westward extension was added when oil was discovered in the Paradon structure. By May 1977, the Reforma Trend extended nearly 50 miles from south to northeast and 34 miles from east to west.[10] The announcement of ten new oil and gas finds in June 1978 resulted in a further extension of 2,320 square miles in Reforma's southern bounds.[11] Since several promising, but as yet undrilled, structures are known to lie outside Reforma's present boundaries, further geographical expansion seems sure (see Map 3.2).

The two facts that Reforma's oil-bearing strata are thick and that its reservoir rocks have a high porosity and permeability explain why the average well produces nearly 6,000 barrels per day.[12] In turn, the coupling of highly productive wells with Reforma's relatively accessible location and nonhostile drilling conditions implies that the lead time between a decision to expand output and the date additional output can come onstream is only two to three years—a lead time far shorter than the time needed to expand output from either the Alaskan North Slope or the North Sea.

Perhaps the most up-to-date summary of Reforma's current status appeared in the *Oil and Gas Journal's* June 1978 survey of Latin American petroleum developments.[13] The *Oil and Gas Journal* noted six important characteristics of Reforma:

1. Current hydrocarbon production from only three structures exceeds 900,000 barrels per day.

2. Some 18 additional structures are now being developed, and most would seem to be at least as large as the three already in production.

3. Reforma's production of crude oil is expected to increase by 500,000 barrels per day in 1978.

4. Most of the early Reforma discoveries have turned out to be far larger than initially expected. For example, in two years of drilling, Pemex has yet to find the boundaries of the Bermudez field. And further drilling has revealed that the Mundo Nuevo structure, labeled a gas field when discovered in 1976, is, in fact, a giant oil field.

5. All structures drilled to date in Reforma have been capable of producing commercial quantities of petroleum.

---

*The gently sloping nature of the structures results in uniform oil-bearing strata, which are much thicker toward the flanks than originally thought. Consequently, waterflooding prospects appear outstanding, and Pemex believes a recovery factor of at least 46 percent can be expected. (See Alvaro Franco, "Giant New Trend Balloons SE Mexico's Oil Potential," *Oil and Gas Journal*, September 19, 1977, p. 83).

Map 3.2: Reforma's Main Structures and Fields

GULF OF MEXICO

BAY OF CAMPECHE

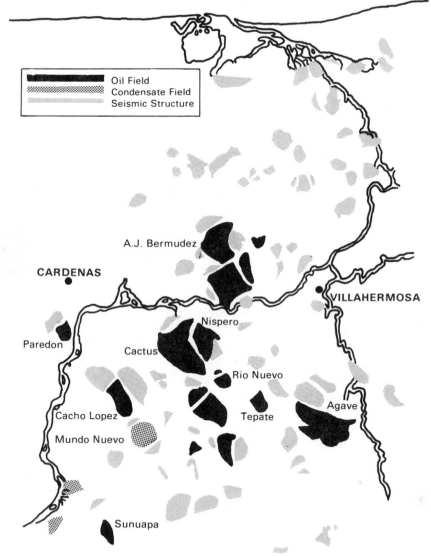

Oil Field
Condensate Field
Seismic Structure

A.J. Bermudez

CARDENAS

VILLAHERMOSA

Nispero

Paredon

Cactus

Rio Nuevo

Cacho Lopez

Tepate

Agave

Mundo Nuevo

Sunuapa

Source: Oil and Gas Journal, June 5, 1978.

6. Pemex's director-general has stated that Reforma will be able to produce at least 3.5 million barrels per day when fully developed. If Pemex maintains its desired 20:1 ration of proved reserves to annual production, Reforma's proved reserves should be at least 25 billion barrels when fully developed.

Campeche

Based on six years of exploration, Pemex has acquired considerable knowledge about Reforma. The same claim cannot yet be made about the recently drilled offshore waters of Campeche. Based on seismic data suggesting that Reforma's structures extended into the Bay of Campeche, Pemex began drilling in Campeche's near-coastal waters in 1974. When Pemex successfully completed Campeche's first three discovery wells in 1975, it encountered enormous oil-bearing structures with physical properties quite similar to those of Reforma's. Thus, it was assumed initially that Reforma's offshore prolongation had been discovered. Based on this evidence, Pemex's director-general stated: "The possible linkings of these areas has aroused great expectations of the existence of reserves comparable to those of some fields in the Middle East."[14] This assumption was modified in mid-1977, when Pemex announced that new interpretations of the seismic data indicated that the Campeche discoveries were not an offshore extension of the Reforma Trend but part of a potentially huge new trend. This new interpretation received further support from the fact that some of Campeche's productive reserves originated in the Paleocene age—a geological strata that is not productive in Reforma. On the basis of seismic studies, Pemex now speculates (mid-1978) that the new Campeche Trend could extend at least 214 miles. This reevaluation prompted Pemex's southern zone manager to say: "We seem to be facing another startling development. It's like the nurse coming out of the delivery room to tell the anxious man he's the proud father of a baby . . . only to come back a while later to tell him it's twins."[15]

According to its director-general, Pemex has already mapped over 200 seismic structures in the Gulf of Campeche (see Map 3.3), "all of them with surprisingly gentle slopes, and thus quite larger than those of Reforma. Should they be oil-bearing, they would dwarf the potential of Chiapas-Tabasco [that is, Reforma]."[16] As of mid-1978, Pemex has already found oil in four Campeche structures: all "are fields of the first magnitude, showing huge oil accumulations in highly-fractured limestones."[17] Another feature of Campeche is that in addition to the oil known to be in the Paleocene strata, "prospects for flush Cretaceous and Jurassic production . . . are excellent."[18] Nevertheless, three cautionary notes should be mentioned. First, available public information about

Map 3.3: Campeche's Main Structures and Fields

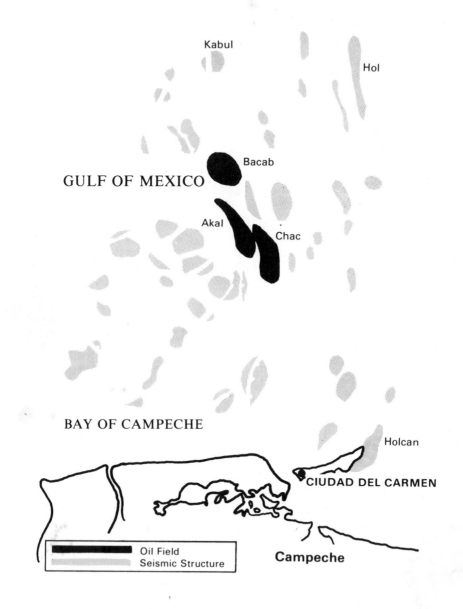

Source: *Oil and Gas Journal*, June 5, 1978.

Campeche is extremely sparse, emanates almost exclusively from Pemex, and is based on only minimal drilling. Therefore, all projections are highly speculative. Second, while the first Campeche well also encountered oil in the Cretaceious and Jurassic zones (the zones that are productive in Reforma) in addition to the Paleocene, water invasion caused the well to be plugged.[19] Thus, though production from the Jurassic and Cretaceous strata is said to be an imminent possibility, none had been reported by September 1978. Third, because of it offshore location, the lead time for expanding Campeche's output is likely to be four to six years; that is, nearly twice as long as the lead time to expand production from the onshore fields of Reforma.

Southeastern Mexico

On the basis of only the sketchy production and reserve data already known about Reforma and Campeche, Mexico's official estimates of its potentially recoverable petroleum reserves were about 200 billion barrels.[20] In short, it already appears that southeastern Mexico's petroleum potential is only exceeded by that of the Persian Gulf. But in sharp contrast to the Persian Gulf, southeastern Mexico is in its very first years as an oil producer. Indeed, because this region's first sizable oil discoveries were made only in 1972—as a direct result of the region's first deep drilling—it will take at least until the mid-1980s before Mexico can regain her former position (held from 1910 to 1920) as one of the world's foremost oil producers and exporters. The *Oil and Gas Journal* has summarized the enormous future potential of southeastern Mexico's burgeoning oil play:

> As exploration advances in southeastern Mexico, Pemex is finding more and more support for its belief that the Chiapas-Tabasco–Gulf of Campeche fields are along a small portion of a giant barrier reef [see Map 3.4] formed along the ancient Yucatan platform during the Cretaceous-Jurassic period.
>
> As pictured by geologists today, the huge atoll-type reef feature extends from some 200 miles west of Reforma (Papaloapan basin) to the open sea in the Gulf of Campeche, and circles to the present peninsula. It then penetrates British Honduras and Guatemala, advances westward into Mexican territory, and closes the loop in the southern part of the Papaloapan basin.
>
> There Pemex already had found significant Cretaceous production—and the highly porous calcerous pays correlate with those of Reforma and Campeche. Best accumulations, Pemex believes, will be found against the barrier reef.[21]

Map 3.4: Pemex's Conception of a Huge Barrier Reef

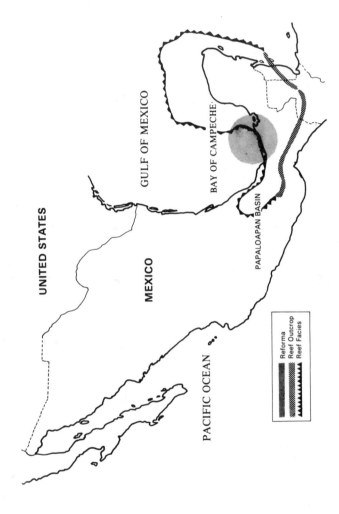

UNITED STATES

MEXICO

GULF OF MEXICO

BAY OF CAMPECHE

PAPALOAPAN BASIN

PACIFIC OCEAN

Reforma
Reef Outcrop
Reef Facies

Source: Oil and Gas Journal, June 5, 1978.

If further exploratory drilling confirms these speculations, southeastern Mexico's petroleum reserves could be well in excess of 200 billion barrels. In short, there is beginning to be some basis for inferring that Mexico may actually have sufficient reserves to supplant Saudi Arabia as the world's richest oil-producing country.

ESTIMATES OF THE TOTAL COST OF PRODUCING SOUTHEASTERN MEXICO'S CRUDE OIL

As a result of fortuitous geological and geographical differences, there are huge regional differences in the total expenditures necessary to discover, develop, and, ultimately, to produce a barrel of crude oil. In mid-1978, total resource costs (that is, all of the unavoidable costs of discovery, development, and production, exclusive of rents, royalties, taxes, and the like) ranged from $.35 to $.50 per barrel in Saudi Arabia, to as high as $6.50 per barrel in the most costly North Sea fields, to $13-plus per barrel for oil from the highest cost stripper wells (that is, wells producing less than ten barrels per day) in the United States.[22] The lowest-cost oil is found in those regions blessed with both relatively low well-completion costs and relatively high productivity.

Well productivity is the most important factor explaining differences in per barrel resource costs. The productivity of the world's currently producing wells ranges from less than one barrel per day to more than 50,000 barrels per day. There are four principal determinants of a well's productivity:

1. The size (that is, amount of oil in place) and thickness of the oil-bearing structure. Wells draining larger and thicker structures typically enjoy higher productivity.

2. The pressure within the oil-bearing structure and the ease or difficulty of forestalling pressure declines as production depletes the field's oil and natural gas. The higher the reservoir pressure, the greater the rate at which its oil is driven to the surface. Pressure declines tend to be slowest when the oil-bearing strata abuts a large underground water aquafier, because as the field's oil and gas are depleted, the water presses in to fill any voids. Pressure maintenance tends to be easiest in fields that are ammenable to waterflooding—the injection of nearby water into the oil-bearing reservoir.

3. The permeability and porosity of the oil reservoir. Productivity is higher when reservoir sediments are porous and highly fractured. This facilitates both a more rapid flow of oil into the well and allows a single well to drain a larger volume of the structure.

4. The viscosity of the oil. The thinner the oil, the faster it will flow into the well hole.

Wells tend to be more costly to complete when the oil-bearing structure is located either in desolate, harsh climates, such as Alaska's North Slope, or in rough and/or deep offshore waters, such as the North Sea. In addition, well costs typically rise more than proportionately with the depth of the oil-bearing strata.

Characteristics Affecting the Cost of Producing Southeastern Mexico's Hydrocarbon Reserves

Southeastern Mexico's hydrocarbon reserves appear to be located in several hundred large geological structures. The region would enjoy considerably lower resource costs if most of its reserves were, instead, consolidated in one or two enormous oil fields: fewer wells would be needed, capital outlays for gas separation and pressure maintenance would be lower, and the costs of delimiting the boundaries of the oil-producing areas would be reduced.

The Reforma reservoirs are subject to rather rapid declines in their natural pressures. Hence, for efficient production, some type of pressure maintenance should be initiated near the beginning of production. This raises production costs. However, Mexico is fortunate that conditions are very amenable for implementing either waterflooding or gas repressuring.[23]

Finally, in addition to the fact that Reforma's oil reserves are relatively deep, two geological factors raise drilling costs. First, there is a thick transition interval of occluded water at high pressures lying above the oil-bearing zones. This has caused cave-ins, brine intrusion, and mud gassification, which, in turn, has slowed drilling and sometimes required redrilling.[24] Second, Pemex has encountered substantial amounts of hydrogen sulfide in several fields. This corrodes the steel well casings, which is alleviated by doubling their thickness.

Compared to the North Sea, Reforma's and Campeche's oil-bearing structures appear to be considerably larger and more numerous and, from the point of view of an oil producer, are located in geographical and geological environments that are far more congenial for production. Therefore, the total resource costs (in 1978 dollars) of producing the average barrel of southeastern Mexico's oil should be substantially less than the $6.50 per barrel cost of (the highest cost) North Sea oil presently being produced. Conversely, because most of Saudi Arabia's oil comes from onshore fields that are considerably larger, shallower, and with wells approximately twice as productive as those already in production at

Reforma and Campeche, the resource costs of Mexico's new found oil must be substantially higher than Saudi Arabia's $.35 to $.50 per barrel.

The remainder of this chapter attempts to glean more precise cost estimates from the publicly available data. Based on conservative assumptions, it concludes that the average resource costs of producing Reforma's and Campeche's oil would not exceed $2.50 per barrel. However, a more plausible point estimate for these costs is about $1.60 per barrel. Within this giant region, costs will, of course, vary substantially around the average. In particular, it seems likely that the extra expense associated with offshore drilling will more than offset the higher productivity presently expected from Campeche's offshore wells.

Estimates of the Resource Costs of Producing Southeastern Mexico's Crude Oil

Five types of data are necessary for calculating rough estimates of the average cost of producing a region's crude oil: (1) the total expenditures necessary to develop the oil field's productive capacity (denoted by D) and the cost of operating the field once it has been developed (denoted by O); (2) the cost of capital (denoted by r_1), that is, the rate of return investors require before they will undertake the capital investment necessary to develop the oil field; (3) the field's initial daily production (denoted by Q); (4) the rate at which production from the typical oil field declines (denoted by r_2) due to the loss of reservoir pressure as its oil and gas is depleted; and (5) the productive life of the region's typical oil field (denoted by n).[25] Given this information, the average resource cost of producing a region's crude oil (denoted by C) can be estimated by solving the equation:

$$C = \frac{(D + O)/Q}{(1 - (1 + r_1 + r_2)^{-n})/(r_1 + r_2)}$$

The numerator expresses the total expenditures per daily barrel of initial production; the denominator expresses the value of an annuity that declines at an annual rate of $r_1 + r_2$ percent and has zero value at the end of n years, when oil production is assumed to cease.

Assuming the expenditures per daily barrel of initial production is given, the resource cost of a region's crude oil will be lower (1) the lower the assumed decline rate and capital cost (that is, the lower the sum of r_1 and r_2), and (2) the longer the field is likely to remain productive (that is, the larger n's value). A plausible lower boundary for Pemex's capital costs (that is r_1) is the approximately 10 percent interest it pays to borrow on international financial markets (see Chapter 4); since Pemex is a national oil company that does not pay taxes but has enormous oil reserves, 15 percent would be a plausible upper boundary on its capital costs—a return that is slightly in excess of the after-tax return on equity presently earned by the large international oil companies. Because of the large size of the Reforma and Campeche oil reserves and Pemex's conservative production practices, the typical field will certainly remain productive for longer than 20 years, and the annual decline rate should be less than 5 percent.

Plausible estimates of southeastern Mexico's expenditures per daily barrel of initial production can be gleaned both from historical and projected data. Between 1973, when Reforma's oil first came on-stream, and year-end 1976 (the last year for which adequate expenditure data are available), Pemex's daily output rose from a 1973 yearly average of 525,000 barrels per day to 960,000 barrels per day.[26] Over the same period, its total expenditures for developing and producing already discovered oil were $1.255 billion.[27] The first column of Table 3.1 shows the average per barrel resource costs that are implied by these historical data. These "historical" cost estimates are subject to a variety of biases.

Because it takes no account of subsequent inflation, the historical data tend to understate mid-1978 costs by as much as 25 to 30 percent. However, this downward bias because of inflation is probably offset by five other factors. First, as time passes, more is known about the region's basic geology. This, in turn, ought to lead to reduced production costs. Second, the expenditure data covers all of Mexico—not just the prolific southeastern regions. The typical Reforma well is at least 30 times as productive as the typical well in the rest of Mexico. Third, the calculation takes no account of the fact that because of the lead time necessary to bring new production on-stream, some of the post-1976 rise in output must be attributed to earlier development expenditures. Fourth, the calculation assumes that the net increase in Mexico's total production equals the net increase in production from southeastern Mexico. In fact, production in the rest of Mexico declined slightly (about 10,000 barrels per day) over this period.[28] Thus, southeastern Mexico's oil production gain is understated. Fifth, because southeastern Mexico's crude oil production was expanding rapidly throughout 1973, the use of 1973 average production causes the production increase to be understated.

In early 1977, Pemex released projections of its total expenditures for drilling and production from 1977 through 1982. Over the same period, it projected Mexico's crude oil output would rise from 953,000 barrels per day to 2.24 million barrels per day. Unfortunately, resource cost estimates cannot be calculated directly from Pemex's projected data because drilling expenditures were not subdivided into those for exploration and those for development. As M. A. Adelman has explained, exploration

TABLE 3.1

Estimates of the Average per Barrel Resource Cost of Producing Southeastern Mexico's Crude Oil

Assumption about Annuity	Pemex's Historical Crude Oil Cost Data	Pemex's Projected Cost Data	
		Crude Oil	Crude Oil and Natural Gas
Case 1:			
r_1 = 10 percent	$1.26	$1.94	$1.51
r_2 = 5 percent			
N = 20 years			
Case 2:			
r_1 = 15 percent	1.62	2.50	1.95
r_2 = 5 percent			
N = 20 years			

Source: Compiled by the author.

expenditures should be excluded from resource cost estimates.[29] From 1972 through 1976, an average of 57 percent of Pemex's total drilling costs were attributable to development.[30] To be conservative (that is, to guard against underestimating resource costs) column 2 of Table 3.1 is based on the assumption that Pemex's projections assume that the share of total costs attributable to development drilling would rise to about 65 percent between 1977 and 1982.

The resource cost estimates deduced from Pemex's projected data do take account of expected inflation. They are, however, subject to three serious types of upward bias. First, the expenditures will be made throughout Mexico, not just in the prolific southeastern region. Second, no account is taken of the fact that Pemex expects that southeastern Mexico's natural gas production will rise by about 2 billion cubic feet per day. Column 3 of Table 3.1 adjusts for this bias on the assumption that a 2 billion cubic feet per day increase in natural gas production is equivalent to an additional increase in crude oil production of 285,000 barrels per day. Third, Pemex subsequently announced (March 1978) that its 1982 production targets would be reached by year-end 1980.

SUMMARY

Southeastern Mexico appears to have well over 100 billion barrels of petroleum producible at relatively low resource costs, which range between $1 and $2.50 per barrel. Assuming Pemex continues to try to maintain a 20:1 ratio of developed petroleum reserves to annual production, reserves of this magnitude would enable the Mexican oil industry to produce, at a minimum, an enormous 13.5 million barrels per day of crude oil and natural gas equivalents. Over the next few years, the key practical problem facing Pemex will be how rapidly it should increase production. Pemex plans to increase oil production by 500,000 barrels per day in 1978. Given Mexico's already known petroleum reserves, even larger year-to-year gains in oil production are technically feasible throughout the 1980s. If Pemex continues to mobilize sufficient resources to continue expanding production at 1978 rates, Mexico's petroleum production will soar to 4.7 million barrels per day in 1985 and 7.2 million barrels per day in 1990.° Such production gains would catapult Mexico into the first ranks of the world's oil exporters. The next chapter assesses whether Pemex has both the will and the talents to meet this challenge.

°These estimates are deliberately conservative. The *Oil and Gas Journal* reports that L. F. Davis, vice-chairman of Atlantic Richfield stated at the annual technical meeting of the Society of Petroleum Engineers (held in Houston, Texas, October 1978) that Arco had increased its long-range forecast for Mexican oil production to 10 million barrels per day by 1992. See "Pemex Insist on Marxist Price for Gas, "*Oil and Gas Journal*, October 9, 1978, p. 36.

NOTES

1. "Mexico Expects U.S. to Buy Its Natural Gas Despite High Price," *Wall Street Journal*, March 12, 1978, p. 8; "Mexico's Oil, Foreign-Exchange Reserves Soar; President Cites Economic Recovery," *Wall Street Journal*, September 8, 1978, p. 6.

2. Ibid. The United States presently produces nearly 9 million barrels per day from its 30 billion barrels of proved reserves.

3. A. A. Meyerhoff and A. E. L. Morris, "Central American Petroleum Potential Centered Mostly in Mexico," *Oil and Gas Journal*, October 17, 1977, p. 109; "Mexico—International Outlook," *World Oil*, August 15, 1977, p. 64.

4. Alvaro Franco, "Bay of Campeche May Rival Reforma Area," *Offshore*, January 1978, p. 44; Meyerhoff and Morris, "Mexico—International Outlook," p. 8.

5. "Pemex Has New Chiapas-Tabasco Finds," *Oil and Gas Journal*, May 2, 1977, p. 64.

6. "Mexico Expects U.S. to Buy Its Natural Gas Despite High Price," op. cit., p. 8.

7. "Mexico Eyes Starved U.S. as Output for Surplus," *Oil and Gas Journal*, June 27, 1977, p. 63.

8. Franco, "Bay of Campeche May Rival Reforma Area," p. 44.

9. Alvaro Franco, "Mexico's Crude-Exporting Role May Be Short-Lived," *Oil and Gas Journal*, May 26, 1975, p. 27.

10. "Pemex Has New Chiapas-Tabasco Finds," p. 120.

11. *Oil and Gas Journal*, June 5, 1978, unnumbered page.

12. Alvaro Franco, "Latin America's Petroleum Surge Gathers Momentum," *Oil and Gas Journal*, June 5, 1978, p. 69.

13. Ibid., pp. 68–70.

14. Alvaro Franco, "Pemex Sees Reforma Extension Offshore," *Oil and Gas Journal*, March 7, 1977, p. 78.

15. Franco, "Bay of Campeche May Rival Reforma Area," p. 44.

16. Franco, "Latin America's Petroleum Surge Gathers Momentum," p. 70.

17. Ibid.

18. Ibid.

19. Franco, "Bay of Campeche May Rival Reforma Area," p. 44.

20. "Mexico's Oil, Foreign-Exchange Reserves Soar; President cites Economic Recovery," *Wall Street Journal* op. cit. p. 6.

21. Franco, "Latin America's Petroleum Surge Gathers Momentum," p. 70.

22. Peter Nulty, "When We'll Start Running Out of Oil," *Fortune*, October 1977, pp. 247–48.

23. Alvaro Franco, "Giant New Trend Balloons SE Mexico's Oil Potential," *Oil and Gas Journal*, September 19, 1977, pp. 83–84.

24. Alvaro Franco, "Pemex Optimizes Reforma Operations," *Oil and Gas Journal*, March 28, 1977, pp. 135–36.

25. See M. A. Adelman, *The World Petroleum Market* (Baltimore, Md.: Johns Hopkins Press, 1972), pp. 45–77.

26. "World Oil Production," *Petroleum Economist*, April 1978, p. 172.

27. "The Future for Mexico," *Euromoney* (supp.), April 1978, p. 18.

28. Ibid.

29. Adelman, op. cit., pp. 73–77. Also, see M. A. Adelman, ed., *Alaskan Oil: Cost and Supply* (New York: Praeger, 1971), p. 21, where he writes that cost "is that amount which would just repay operating cost plus a barely sufficient return on the needed development investment. . . . The excess of price over cost is a return for exploration risk, over and above development risk. If the excess is small, that means exploration is not worthwhile."

30. The drilling expenditure figures are from "The Future for Mexico," op. cit., p. 18. During the period 1972–76, drilling expenditures are broken down between exploration and development, the latter accounting for 57 percent of total drilling costs. Drilling costs are not broken down for the 1976–82 period.

4

PEMEX

Foreign participation in Mexico's oil industry came to an abrupt halt in March 1938 when the Mexican government expropriated nearly all foreign holdings. Literally overnight, control over the industry passed from the hands of U.S. and British oil companies and into the hands of the Mexican government. Although the government's motives for expropriation were primarily political, the initial consequences were largely economic and technical: the removal from the oil industry of foreign capital and expertise left a vacuum the Mexican government was forced to fill. This chapter describes how Mexico's national oil company, Petróleos Méxicanos, or Pemex, has performed in this role and assesses the likelihood of its being able to meet the challenge of rapidly developing Mexico's giant petroleum reserves.

BACKGROUND

Pemex did not rise phoenixlike from the ashes of the foreign oil companies expropriated in March 1938. Rather, in 1925, the Mexican government, perhaps in anticipation of a confrontation with the foreign oil interests, had chartered a public agency—the National Petroleum Administration (NPA)—to engage in production and refining operations in competition with the foreign private companies. In 1934, the NPA's

Kenneth Colli and Jamie Kirkpatrick helped to co-author this chapter. Most of the information about Pemex's current constraints and the extent of present involvement of foreign companies in Mexico's oil play has been gleaned from very sketchy press reports. Because of the rapidly changing nature of the Mexican oil play and the secretiveness of the involved parties, we cannot vouch for their accuracy.

functions were assumed by a semiprivate organization, Petróleo de México (Petrómex), in which the government barred all foreign stock ownership and reserved a minimum of 40 percent ownership for itself. Though small, Petrómex was a fully integrated oil company.

As the day of expropriation approached, the Mexican government sought a fuller measure of control over the domestic oil industry. It chose to supplement Petrómex with a new government department, the General Administration of National Petroleum. At the time of nationalization, this all-Mexican operation was producing 16,000 barrels per day.[1] More importantly, it provided the government with a pool of trained personnel, who along with the few Mexicans in positions of management in the foreign companies were to direct Mexico's petroleum industry.

Mexico's immediate task after nationalization was to maintain the productivity of the fields vacated by the foreign operators. The necessary infrastructure was present, since all capital goods remained in place and had become the property of the nation. However, much of the equipment was outdated or in poor condition, owing to the foreign companies' reluctance to invest in maintaining and updating facilities thought likely to be expropriated. In April 1938, a month after expropriation, production declined by almost 50 percent. However, within six months, it had nearly regained the preexpropriation level—despite the hostility of the expropriated companies, who had instituted a boycott of exported Mexican petroleum and related products and who were also putting pressure on U.S. suppliers to sever contracts with Mexico. The government was able to begin building a wholly national production capacity with only a temporary sharp drop in output from preexpropriation levels—primarily because the populace, especially the petroleum workers, were so supportive of expropriation.

The government faced strong economic and financial pressures to maintain Mexico's oil production. In order to ensure that expropriation would not violate international law, President Cárdenas had promised full compensation for all expropriated assets. The entire Mexican economy needed to be mobilized to meet the demands of servicing this nearly $200 million debt. Moreover, because petroleum production was such an important element in the domestic economy, Mexico could ill afford either reduced oil revenues or the need to shift resources from other sectors of the economy in order to support a failing petroleum industry. Finally, maintenance of production was also important for protecting Mexico's national self-esteem. Mexico had acted strongly, many said unwisely, in nationalizing its oil industry. To discredit the skeptics, the government needed to demonstrate that Mexicans could, indeed, assume full responsibility for an entire industry and, thus, make economic nationalism more than a political platitude.

In order to effect these goals, the government reorganized the petroleum industry on June 7, 1938: President Cárdenas invested two public agencies, jointly controlled by labor and government, with full operational responsibility for the industry. Pemex would handle all phases of the industry up to marketing, at which point the Distribuidora de Petróleos Méxicanos would take over. The resulting diffusion of responsibility led to duplication of effort and wasted expenditures. What was even more harmful was that it proved impossible to control the power of the local labor unions. By early 1940, the strains between Pemex and the Distribuidora, on the one hand, and between labor and the government agencies, on the other hand, were so debilitating that President Cárdenas ordered abolishment of the Distribuidora and transferred its assets and functions to Pemex. Once centralization of administration was accomplished, Pemex quickly coordinated the local labor councils and established its position as the controlling body of the Mexican oil industry.

Although Pemex has evolved over the years to adapt to changing conditions in domestic and international markets, the company has continued to be structured along the lines established in the 1940 reorganization. The board of directors has nine members: four representing Mexico's national oil workers' union, the Sindicato de Trabajadores Petróleros de la República (STPRM), and five appointed by the president and representing the government. Mexico's president also appoints Pemex's director-general, the official responsible for overseeing all operations of the company.

By statute, Pemex's principal aim is not profit but the achievement of social goals, such as a high level of employment and the provision of petroleum to the domestic economy at low and stable prices. Thus, from Pemex's inception, the government has always enforced strict price controls on domestic oil sales—Mexicans were paying only about $6 per barrel for oil in early 1978. Nevertheless, because the Mexican government believed the private foreign oil companies had reaped excessive profits from their Mexican investments, Pemex has always felt some responsibility not only to cover its own costs but to make a profit for the government. Pemex collects petroleum consumption taxes and export taxes for the nation. In addition, it has been required to pay a percentage of its gross income to the government as royalties.[2]

Because of its key role at the creation of Mexico's national oil industry, STPRM has always enjoyed a unique position vis-a-vis Pemex. Labor has insisted that Pemex should be exclusively devoted to promoting social welfare—especially, the welfare of its workers. Pemex had more than 98,000 employees in 1977, and its manpower-to-production ratio was nearly three times higher than that of highly integrated U.S. companies. Skilled labor is in short supply in Mexico, and STPRM's workers are among the highest paid; in addition, they enjoy substantial subsidies for education, medical services, housing, and recreation. The fact that more than half of all workers hired by Pemex in 1976 are thought to have been related to other Pemex employees suggests that the union has been successful at "feathering" its nest.

Throughout Pemex's 40-year history, labor and management have held opposing visions of Pemex's role. STPRM's consistent efforts to increase labor's benefits has made it much more difficult for management to run a financially sound enterprise. J. Paul Getty is reputed to have said: "Pemex is the only oil company I know that ever lost money." Though the quotation may be apocryphal, it does express a widely held view of private oil companies and vendors of products and services for the international oil business. Pemex has always had an overstaffed, highly paid labor force, and it has been suggested that its managers frequently make sure their friends are the recipients of lucrative contracts (to provide the company with oil field services and supplies). The *Economist* wrote in April 1978:

> PEMEX has a bad reputation that it has to live down. Politicians used it as a dumping ground for those to whom favors were owed. It used to be the custom to offer politicians' wives the "concession" on a brace of PEMEX filling stations. The oil union is immensely powerful and sells jobs in the company to those that want to work there.[3]

A further problem arises because Pemex has a rigid hierarchical structure that concentrates all decision-making responsibility in the hands of a few top officials. According to *Fortune*, "no decisions except those of a technical nature are made below the level of departmental manager—a group of only a few dozen people in a corporation with 98,000 employees. As with a dinosaur, the immense and rather lumbering body is kept in action by a diligent little head cropping away day and night."[4] Pemex has also encountered problems in trying to coordinate planning among the five directorates into which the company is divided. Each directorate operates as a small enclave and guards against the encroachment of its responsibilities by sister divisions (see Figure 4.1).

There is little doubt that Pemex has far higher costs than the private multinational oil companies. Nevertheless, the frequently heard negative evaluations of its talents and accomplishments are probably too harsh. As the first non-Communist national oil company, Pemex has always been a pioneer, chartered to advance social and national goals in addition to economic efficiency. Even today, it is the only national oil company that really is a vertically integrated firm—the others subcontract nearly all difficult tasks to private, typically foreign, firms. Moreover, it is the only national oil company ever to find substantial new quantities of oil in a Third World country. Because of these successes, Pemex is (and probably deserves to be) regarded as a role model throughout the developing world.

Figure 4.1: Pemex Organization Chart

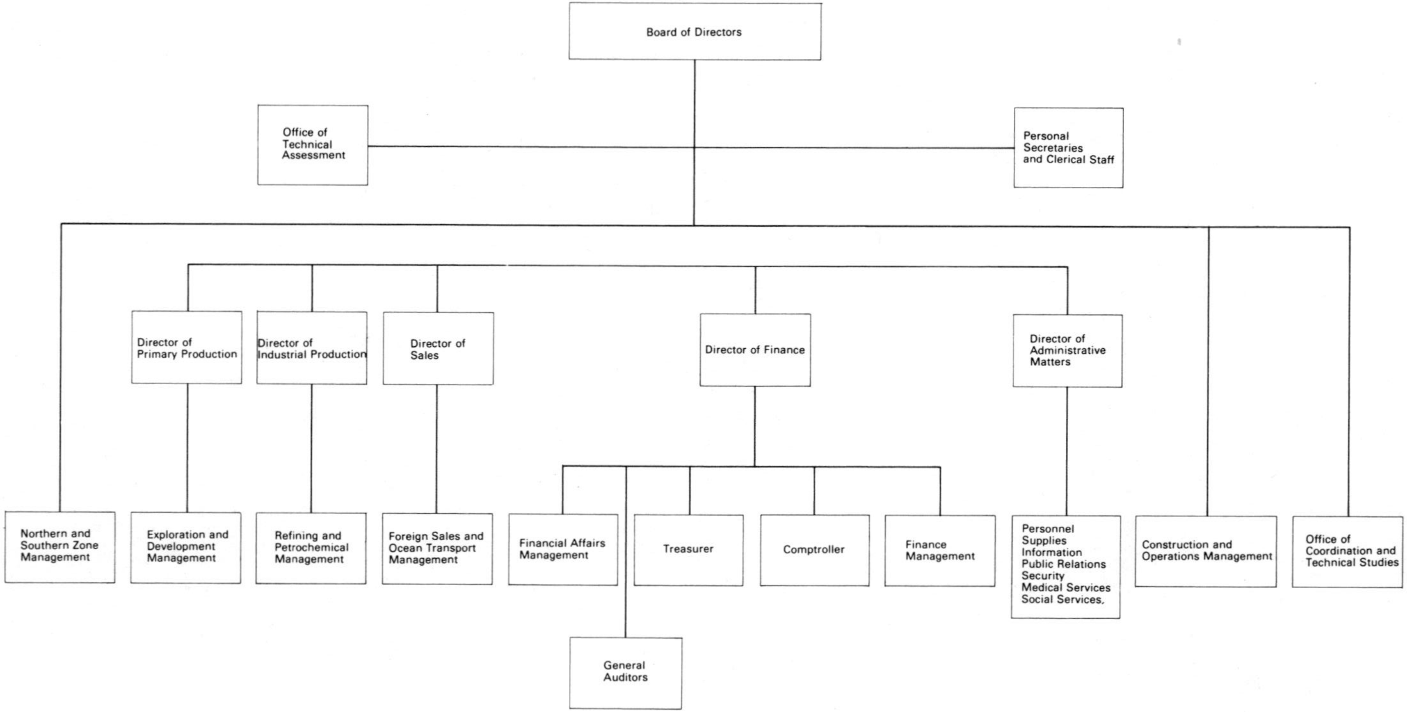

Source: James Kirkpatrick, "Foreign Participation in the Mexican Oil Industry." MALD Thesis, Fletcher School of Law and Diplomacy, May 1978.

CONSTRAINTS ON PRODUCTION

Pemex's director-general, Jorge Díaz Serrano, has said that Pemex will "produce all the oil we can now rather than later when we thus would regret not having lived up to this historic moment."[5] The fact that the government has authorized Pemex to spend $6.3 billion of the $10.8 billion earmarked for all Mexican industry in 1978 confirms that Serrano's statement represents national policy and not merely rhetorical excess.[6] Table 4.1 lists Pemex's actual capital expenditures from 1972 through 1976; Table 4.2 lists its planned capital expenditures from 1977 through 1982. The goals when the 1977–82 capital program was introduced were the following: (1) to drill 1,324 wildcats and 2,152 development wells; (2) to increase production from less than 1 million barrels per day to 2.2 million barrels per day; (3) to increase exports of crude oil from 200,000 barrels per day to 1.1 million barrels per day and to increase exports of refined petroleum products from virtually nothing in 1977 to 300,000 barrels per day; (4) to make Mexico self-sufficient in petrochemicals by tripling petrochemical capacity to 18.6 million tons; and (5) to construct a 48-inch natural gas pipeline from Tabasco to Monterrey with a branch to the U.S. border.[7]

At the time of their announcement, these goals were thought to be ambitious. Yet, due to the enormous reserves of oil at Reforma and Campeche, it soon became apparent that the targets for production and export of crude oil could be exceeded easily. As Chapter 3 notes, in March 1978, Pemex revealed that it now expected to reach its 1982 production and export targets for crude oil by year-end 1980.

Based on its post-1974 track record and the enormous, ever-growing petroleum reserves in southeastern Mexico, Pemex certainly appears capable of raising its petroleum output to at least 4.7 million barrels per day in 1985 and to 7.2 million barrels per day in 1990 (that is, by annual increments of 500,000 barrels per day—Pemex's rate of expansion in 1978). But as Chapter 3 explains, there is already evidence that Mexican petroleum reserves are sufficient to support a petroleum industry that could produce at least 13.5 million barrels per day of crude oil and natural gas. In short, accepting at face value Serrano's statement about Pemex's goal of producing all the oil it can now rather than later, what are the plausible limits on Pemex's petroleum production during the 1980s? The remainder of this chapter concludes that no definitive answers to this question are possible, because it depends on the extent to which Pemex allows foreign participation to aid it in overcoming its two principal near-term constraints: shortages of financial reserves and inadequate technical capabilities and trained personnel.

TABLE 4.1

Pemex's Capital Expenditures: 1972–76
(millions of dollars)

Item	1972	1973	1974	1975	1976
Exploration	$ 38.712	$ 39.200	$ 46.072	$ 72.000	$ 112.411
Drilling:					
Exploration	98.640	88.632	110.065	122.640	130.188
Development	99.552	116.248	148.583	183.864	191.412
Production	46.760	96.072	118.097	144.801	256.284
Refining	44.936	62.288	96.564	259.217	270.064
Petrochemicals	44.576	77.944	96.093	145.990	258.081
Transportation and marketing	90.400	136.712	94.773	163.936	22.038
Other	5.280	11.472	10.763	2.898	3.212
Total	486.856	628.568	721.010	1,095.346	1,443.690

Source: "The Future for Mexico," Euromoney (supp.), April 1978, p. 18.

TABLE 4.2

Pemex's Planned Capital Expenditures: 1977–82
(millions of dollars)

Item	1977	1978	1979	1980	1981	1982
Exploration	$ 92.2	$ 112.0	$ 138.8	$ 175.4	$ 226.6	$ 276.7
Drilling	492.3	528.1	665.9	760.2	756.1	831.2
Production	391.3	626.0	545.1	463.8	521.5	556.2
Refining	701.9	477.4	333.3	294.2	233.8	294.0
Petrochemicals	565.4	662.3	602.1	286.4	166.5	155.9
Transporation and marketing	394.5	400.1	279.1	330.5	282.5	247.3
Other	34.9	40.4	42.7	46.0	48.4	51.4
Total	2,672.5	2,846.3	2,607.0	2,356.5	2,235.4	2,412.7

Source: "The Future for Mexico," *Euromoney* (supp.), April 1978, p. 18.

Pemex's Past Attitudes toward Foreign Participation

Immediately following nationalization, the expropriated oil companies imposed a boycott on purchases of Mexican crude oil: Mexico's petroleum exports fell from 2 million barrels in February 1978 to only 311,000 barrels in April 1938.[8] Though wartime pressures caused the boycott to collapse by 1940, Mexico's oil exports continued to stagnate because of war-caused tanker shortages.

Following World War II, the Mexican government realized that domestic markets were too narrow either to finance or supply the capital equipment necessary to modernize Pemex's increasingly obsolete oil field equipment. The government began to reconsider the nature of the ban it had imposed on foreign participation in the 1938 and subsequent decrees. The problem was an extremely delicate one, because Mexican labor maintained its suspicion of the "imperialists" it had defeated only a decade earlier.

In late 1947, Pemex obtained labor's consent to make an exploration contract with a small U.S. company that had no previous involvement in the Mexican oil industry. Labor appears to have acquiesced in this instance due to the company's small size, which appeared to pose no threat. However, that very virtue proved to be a vice, because the company was unable to obtain the necessary drilling equipment; therefore, the contract was allowed to lapse.

The next attempt to reinstitute foreign participation came in 1949, when Pemex signed drilling contracts with a U.S. group that included the American Independent Oil Company, (AIOC), Signal Oil and Gas, and Edwin W. Pauley. The contractors were to perform drilling services for Pemex and had no rights over any petroleum they might discover. From an operational standpoint, these 1949 drilling contracts proved to be of slight value. By 1958, only 2 percent of Pemex's production came from reserves located and developed by the U.S. drillers. Also, no new drilling technologies were introduced to Mexico. However, the experiment did solidify the government's belief that any foreign participation would have to be within the framework of Mexican law and petroleum policy, that is, that control over production must always reside in Pemex.

Pemex also began to look to foreign capital markets in 1949. The company attempted to negotiate a $500 million loan from the Export-Import Bank. The application was endorsed by the U.S. departments of Defense and State, because they thought development of Mexico's petroleum industry would also promote political and economic security in the western hemisphere. Nevertheless, the loan was refused, because U.S. oil companies succeeded in persuading the Export-Import Bank to adhere to its policy of not displacing opportunities for private capital investment with public credits. The oil companies preferred that private capital, *their* private capital, be the vehicle for further development of Mexico's oil industry. However, choice of this alternative would have meant political suicide to the Mexicans responsible for the loan application.

In the 1950s, the issue of foreign participation surfaced again as technical advances made offshore exploration appear promising. Senator Millard Tydings, representing a group of private U.S. investors, offered to supply the equipment necessary for exploring and producing Mexico's marine deposits. The Tydings group hoped that in return, Mexico might be willing to modify its position on petroleum ownership, at least as it applied to those resources lying beneath the seabed. But the Mexican government saw no valid distinction between the ownership of subsoil and submarine rights: both belonged to the nation. According to Antonio Bermúdez, Pemex's director-general at the time, it "would have been easier to change the colors of the Mexican flag than to change the country's laws relating to petroleum."[9] The offer was dismissed politely.

Pemex's early flirtations with foreign participation came to a close in November 1958 with the passage of a new petroleum regulatory law. This law, once again, clearly reaffirmed that only the nation could exploit Mexico's oil resources and that even concessions were not titles to produce oil. The law also extended the concept of public control over Pemex's downstream domestic operations, such as refining, transportation, and marketing. Private participation was not excluded from downstream operations, but the law clearly stated that Pemex could only award contracts deemed to be in the public interest. In short, throughout the 1960s, Mexico did not offer an attractive business climate for either foreign petroleum companies or foreign sellers of oil field and related services.

Current Constraints on Pemex

Because the great predominance of Pemex's drilling has always been onshore, Mexico's indigenous industrial infrastructure has been geared toward manufacturing oil field goods and services to supply this market. But Mexico's largest petroleum reserves seem likely to be offshore in the Gulf of Campeche. The shift to substantial offshore production will create strong new pressures for Pemex to purchase the necessary equipment and technical expertise from foreign manufacturers.

Rapid expansion of Mexico's offshore oil production will also be hindered by a shortage of skilled labor. Pemex has always prided itself on the competence of its workers. The foreign oil companies were obliged by law to train and utilize Mexican nationals in the operations of their

Mexican subsidiaries. Thus, even in the days immediately following expropriation, skilled labor was one of the few areas in which Pemex did not suffer a severe shortage. But before the massive strikes in Campeche, Mexico had only three offshore rigs.[10] Hence, Pemex's large labor force has had almost no experience in offshore operations. The 1977–82 oil industry development plan calls for drilling exploratory wells in 24 separate offshore structures, and 120 development wells are to be drilled from ten fixed platforms in Campeche's waters.[11] Substantial reliance on foreign equipment and technicians seems inevitable if Pemex is to carry out this part of its plans.

Pemex's other major bottleneck is likely to occur in the development of new production technology for its rapidly expanding oil operations. While Pemex's R&D affiliate, the Mexican Petroleum Institute, has a good record in modifying production technologies to accommodate the specific conditions of Mexico's petroleum sector, it appears to lack the capability to supply the technical needs of an industry as large and rapidly growing as the one Mexico may choose to develop. Mexico will most likely seek access to advanced foreign technology through the purchase of licensing rights.

Future Reliance on Foreign Participation

Now that Mexico is aware of its enormous petroleum potential, it must select from among three strategies for the development of its hydrocarbons:

1. A go-it-alone strategy in which the rate of oil development is set by the rate of expansion of Pemex. Under this strategy, Pemex would maintain total control over every phase of the Mexican oil industry.

2. All-out rapid development—a strategy that would necessitate contracting out production of crude oil (especially offshore), downstream refining, petrochemicals, and transportation to foreign service companies on a large scale.

3. A compromise strategy in which Pemex would retain full control over the industry, but foreign service companies would be employed to overcome specific bottlenecks encountered by Pemex.

The go-it-alone strategy adheres closely to the philosophical principles engrained in the Mexican population since the revolution and guarantees the greatest bargaining power for Pemex's workers. Thus, it is not surprising that the main advocates of this option have been the trade unions and extreme nationalists. However, the dismal economic conditions plaguing Mexico have seriously undercut their position. Although good statistics are not available, most authorities estimate that at least 10 percent of the labor force is unemployed—when underemployment is included, the figure jumps to 40 percent.[12] An even worse crisis may lie in Mexico's near future: Mexico's population is growing at 3.4 percent per year, one of the highest growth rates in the world. Sharp increases in production of hydrocarbons offer the only credible prospect for providing Mexico with resources to help defuse this population bombshell.

The severity of Mexico's economic problems was an important factor behind López Portillo's succession to the presidency from finance minister in 1976. Upon entering office, Portillo made it clear he intended to use petroleum as the key sector for rebuilding the Mexican economy. The president's thoughts were reflected in a speech by Pemex's director-general before the Chamber of Deputies:

> Both domestically and abroad, Mexico will be stronger politically to the extent that it manages to increase the power of its petroleum industry, but we should not forget that this is a race against the clock.
>
> The social costs of our failing to follow a dynamic production policy will be very great. Every industrial development program for primary and secondary petrochemicals and associated manufacturers will be slowed down in the short and medium term. . . .
>
> I cannot think that the alternative is very difficult to see. We need to produce more crude, more refined, more petrochemicals, more liquified gas and more natural methane gas, and we need to trade these products with those who will pay the best price for them, without distinction of ideology or preferences for particular groups. The interests of Mexico demand this.[13]

The urgency expressed in this statement suggests that Mexico wishes to move rapidly forward in developing its hydrocarbon potential. Having made this decision, the Portillo administration is confronted with the issue of determining how much foreign involvement is politically feasible. If there were no domestic constraints, all-out development would be the most attractive alternative. Inviting foreign service companies to participate in the development of Mexico's oil on a large scale would be the quickest way for Mexico to raise its production capacity. But because it would conflict with the spirit of the revolution, such a course does not appear to be feasible. In the words of President Portillo: "Our Constitution reserves the exploitation of hydrocarbons for the nation. It is our obligation to ensure that this provision, which has given our country such substantial benefits, remains in force always."[14]

Faced with enormous demands for oil revenues and strong domestic opposition to foreign participation, the Portillo government has followed a delicate middle course in developing Mexico's oil: Pemex will continue to be assigned full responsibility for running Mexico's oil industry, but foreign contractors will be hired on a project basis to circumvent specific bottlenecks. Presently, it appears that Pemex will continue to perform nearly all activities necessary to develop and produce onshore reserves. In early 1978, it already had about 190 drilling rigs operating in onshore areas (compared with about 2,000 operating in the entire United States), and all were staffed by Pemex crews.[15] In view of Pemex's accomplishments in raising Reforma's production from practically nothing in 1974 to 70 percent of Mexico's total output in 1977, there should be no doubt that it has the talents to handle this assignment.[16]

Recognizing its lack of offshore expertise, Pemex moved in mid-1977 to hire Haliburton's world famous Brown & Root subsidiary to undertake project management of an estimated $500 million effort to organize the engineering work, coordinate purchasing of production platforms, gather lines and pipelines, and build the shore facilities necessary to begin the first large-scale commercial production from the Bay of Campeche.[17] However, Pemex has no intention of relying exclusively on foreign firms to provide offshore expertise. It acquired an additional five offshore rigs in early 1978, and although there will be numerous foreign technical advisers, all are to be manned by Pemex crews.[18] In a sharp departure from past practice, two of these rigs are to be operated by a joint venture of the Texas-based Rowan companies (with a 49 percent equity share) and a group of Mexican private investors.[19] This demonstrates that in areas where Pemex's expertise is limited, it is willing to make an exception to its hard-line ban on direct foreign participation (albeit with guarantees that preserve Mexican ownership and control of the venture). In sum, Pemex appears to believe that by 1982, when Brown & Root is scheduled to complete its work as project managers to develop the first Campeche fields, Pemex will have the ability to undertake the role of project manager for developing Campeche's other fields. However, large numbers of foreign subcontractors are likely to continue to play a substantial part in this effort.

Construction of the giant 48-inch pipeline originally planned to carry natural gas from Tabasco to the Texas border has been another project for which Pemex has chosen to go abroad for assistance. Early in the pipeline's design phase, Pemex contracted with Tenneco Inter America to provide engineering services. In addition, most Mexican construction firms engaged in this project have associated themselves with a U.S. counterpart. Overall engineering, testing, and management are in the hands of a joint venture of ICA, a large Mexican construction firm, and a U.S. company, Bechtel.[20]

Compared with other developing countries, Mexico has long had a well-developed indigenous oil field supply industry. However, this industry's scale is geared much more for oil fields of the size Pemex was developing prior to the giant Reforma discoveries. Thus, in order to meet its development plans, Pemex has also been compelled to purchase large quantities of drilling equipment, production machinery, pipeline steel, valves, turbines, bits, and compressors from the big multinational oil field equipment firms.

The final area of major activity for foreign interests is in the manufacture and processing of petrochemicals. After expenditures for drilling and production, expenditures for petrochemicals account for the largest share of Pemex's planned capital expenditures from 1977 through 1982. Almost $1 billion was spent to expand Mexico's petrochemical facilities in 1977. The largest projects were the expansion of an ammonia plant designed by a U.S. firm, Pullman-Kellog, and construction of a new ethylene facility. Because these downstream activities are somewhat removed from the politically sensitive question of who conducts Mexico's petroleum exploration and development, petrochemicals appear to be an especially fertile ground for increased foreign investment.

Financial Constraints

Pemex's strategy is to develop its own capacity as quickly as possible while relying on foreign service companies to fill what it hopes are temporary shortages of capital, technical expertise, and skilled personnel. An important question associated with this strategy involves financial arrangements—in light of Pemex's historic shortage of cash, how will it finance the rapid development of the Reforma and Campeche fields? The latest six-year plan projects a fivefold increase in investments from the previous six-year period.

In the eyes of both Mexican and international bankers and investors, Pemex's credit is tied closely to the quantity of Mexico's petroleum reserves. Before the enormous Reforma reserves were announced to the world, Mexico was increasingly viewed as a bad credit risk due to overreliance on foreign borrowing to finance internal development. Under President Portillo's predecessor, Luis Echeverría, Mexico's public debt soared above $20 billion. As the debt rose, so did concern among the country's creditors. By 1976, investors had become so chary of Mexican borrowings that the International Monetary Fund (IMF) intervened and imposed a ceiling on annual loans to Mexico of no more than 12 percent of its gross domestic product (GDP)—approximately $3 billion.

One of Portillo's first acts after he assumed power in late 1976 was to create a task force to determine accurately the quantity of Mexico's oil reserves. Within two months, the task force reported that Mexico's proved reserves were in excess of 11 billion barrels, nearly twice the amount Pemex had previously reported. Pemex management then hired the U.S. consulting firm of DeGolyer & McNaughton to audit and, if possible, verify these new estimates. DeGolyer & McNaughton not only verified the task force figures but increased them. Not surprisingly, foreign confidence in Mexico was restored at a rate almost proportional to the increase in the nation's proved reserves. The IMF removed its strict borrowing ceiling in 1977, and Mexico's government has taken steps to slow domestic inflation.

Evidence of the success of Portillo's policies is increasingly visible: in 1977, Pemex secured only one foreign loan—$350 million from a consortium managed by the Chase Manhattan Bank at a rate one and one-half points above the London Interbank rate. In the first four months of 1978 Pemex, secured four separate loans totaling $1.2 billion, the largest of which was a $1 billion loan from a consortium led by the Bank of America and Manufacturers' Hanover at a rate one and one-quarter points above the London rate.[21] It is a good sign of increasing credit worthiness when dollar amounts increase substantially while point spreads decrease. As of April 1978, it was anticipated that Mexico would be the largest single borrower in the Euromarket during 1978.[22]

In addition to direct borrowing on international capital markets, Pemex has two other methods for obtaining the funds necessary to finance its rapid expansion: prepayments from foreign customers and export credits from the home countries of the firms from which Pemex purchases equipment. Pemex has made extensive use of export credits. Pemex flirted with prepayments in 1976–77 after the IMF imposed the $3 billion ceiling on Mexico's foreign borrowing. If foreign customers prepaid for future deliveries of hydrocarbons, the transaction would be entered in the Mexican accounts as a credit for export earnings and not as a debit for borrowing. Thus, Mexico could circumvent the IMF limits.[23] A several hundred million dollar prepayment agreement was negotiated with six U.S. natural gas transmission companies; however, the arrangement was not consummated, because the U.S. government refused to authorize the importation of Mexican gas at the agreed upon price of $2.60 per Mcf. Prepayment loans are more expensive, because the foreign customer has to borrow the money for the prepayment. Hence, this method of financing has been disregarded since the IMF lifted its credit ceiling.

SUMMARY

Pemex today seems to be in a transitional period in its attitude toward foreign participation. While maintaining official abstinence, especially in onshore Reforma, there is a quiet but noticeable trend to reinstate limited foreign participation in offshore operations and in the downstream phases of the oil industry. But the lessons of the past certainly have not been forgotten. Pemex and the government are very careful to preserve Mexico's economic and energy independence by controlling all forms of foreign participation. These controls need to be strict enough to assuage those Mexicans, particularly STPRM, who continue to fear foreign domination but not so tight as to deter the foreign investment and expertise that Mexico needs if it is to live up to its "historic moment."

NOTES

1. J. Richard Powell, *The Mexican Petroleum Industry: 1938-1950* (Berkeley: University of California Press, 1956), p. 35.

2. Fredda Jean Bullard, *Mexico's Natural Gas* (Austin: University of Texas Press, 1968), p. 8.

3. "Survey of Mexico," *The Economist* (supp.), April 22, 1978, p. 24.

4. Hugh Sanderman, "Pemex Comes Out of Its Shell," *Fortune*, April 10, 1978, p. 18.

5. "Mexico Expects U.S. to Buy Its Natural Gas Despite High Price," *Wall Street Journal*, March 21, 1978, p. 18.

6. Patricia Nelson, "Mexico Promoting Industry," *Journal of Commerce*, December 20, 1977, p. 9.

7. "Survey of Mexico," p. 23.

8. Powell, op. cit., pp. 97–98.

9. Antonio J. Bermúdez, *The Mexican National Petroleum Industry: A Case Study in Nationalization* (Palo Alto, Calif.: Institute of Hispanic American and Luso-Brazilian Studies, Stanford University, 1963), p. 35.

10. Alvaro Franco, "Bay of Campeche May Rival Reforma Area," *Offshore*, January 1978, p. 43.

11. Ibid.

12. James L. Schlagheck, *The Political, Economic, and Labor Climate in Mexico* (Philadelphia: Wharton School, University of Pennsylvania, 1977), p. 85.

13. Jorge Díaz Serrano, "Speech before the Chamber of Deputies, October 26, 1977," in *Comercio Exterior de Mexico*, December 1977, p. 483.

14. President Jose Lopez Portillo, First State of the Nation Report (September 1, 1977) in *Comercio Exterior de Mexico*, September 1977, p. 337.

15. Sanderman, op. cit., p. 46.

16. George W. Grayson, "The Oil Boom," *Foreign Policy*, Winter 1977-78, p. 76.

17. "Pemex to Brown & Root: Yankee, Come In," *Forbes*, August 15, 1977, p. 28.

18. Inferred from articles in *Offshore*, January–March 1978.

19. "International Briefs," *Oil and Gas Journal*, February 20, 1978, p. 90.

20. Richard R. Fagen and Henry R. Nau, "Mexican Gas: The Northern Connection" (Paper delivered at Conference on the United States, U.S. Foreign Policy and Latin American and Caribbean Regimes, Washington, D.C., March 27-31, 1978), p. 26.

21. "The Future for Mexico," *Euromoney* (supp.), April 1978, pp. 34–35.

22. Ibid., p. 33.

23. Fagen and Nau, op. cit., p. 27.

II

MEXICO AND THE WORLD OIL TRADE

INTRODUCTION

Southeastern Mexico almost certainly has a resource base of 100-plus billion barrels of petroleum (recoverable), using presently commercial technologies and at costs far below the prices currently prevailing on world markets. Assuming continuation of 1978 expansion plans, Pemex has the capability of producing at least 4.7 million barrels per day of crude oil and natural gas equivalents by 1985 and 7.2 million barrels per day in 1990. Production at these (and potentially much higher) levels would catapult Mexico into the position of the world's second- or third-largest petroleum exporter by the late 1980s. Based on the post-1973 experiences of Saudi Arabia and Iran, it seems likely that Mexico's international status and influence will soar as rapidly as its petroleum exports. The principal issues Part II addresses are (1) what would be the benefits and costs to Mexico of a decision to expand oil production at the high rates that appear to be commercially and technically feasible, and (2) what would be the implications of Mexico's decision for other countries, especially the United States? The prevailing theme—some will argue bias—is that Mexico and the United States would both reap large economic and political gains from the rapid acceleration of Mexico's petroleum exports.

5

MARKETS FOR
MEXICO'S PETROLEUM

The United States is the natural market for Mexican exports of oil and natural gas. The reasons are twofold. First, the United States with its enormous thirst for energy offers a nearly insatiable potential market for Mexico's petroleum. U.S. oil imports (averaging approximately 8 million barrels per day in 1978) will certainly remain high through the next decade—some forecasts suggest they will approach 12 to 13 million barrels per day in 1990. Second, because of the relative proximity of the U.S. Gulf Coast and Texas import facilities to Mexico's petroleum-producing areas, transport costs to the United States will be substantially lower than to any other large world market. In the case of natural gas, these locational economies are so strong that prior to the late 1980s, the United States offers the only commercially viable market for large Mexican exports.* Unlike natural gas, Mexican oil can be transported and sold profitably all over the world. However, Mexico will earn lower per barrel profits on all oil not sold in either the United States or Canada. This chapter attempts to quantify Mexico's foregone profits if, for either domestic or international political considerations, it chooses to sell its oil in other world markets.

WORLD OIL PRICING: PERSIAN GULF PLUS FREIGHT

Crude oil is a highly fungible product that is nearly always refined

---

This chapter was co-authored by Peter Milone.

* To be transported, natural gas must either be shipped through high-pressure pipelines or liquefied and shipped on specially constructed liquefied natural gas (LNG) tankers. Liquefaction is highly capital intensive and would raise the cost of exporting Mexico's natural gas by at least $1.20 per thousand cubic feet. In addition, it would take at least five years to build the necessary facilities.

into fuels or petrochemicals. After making appropriate allowances for differences in chemical or physical characteristics, which affect the yield, cost, and value of the refined products (for example, sulphur content and specific gravity), the price that a barrel of crude oil can fetch in any market is determined by the cost of the marginal crude it replaces. Presently, the Persian Gulf countries are the marginal source of crude oil for all major world markets, and therefore, crude oil's price in any world market is the free-on-board (fob) Persian Gulf price plus all freight and related costs.

Competitive forces entail that assuming appropriate adjustments are made for quality differentials, the delivered price of oil in any given market should be independent of its country of origin. Thus, it is possible to calculate the fob price country j receives from its oil exports to country i (denoted by P_j^i) by solving the equation

$$P_j^i = P_{PG} + F_{PG}^i - F_j^i$$

where P_{PG} is the Persian Gulf reference (or base) price, F_{PG}^i is the freight cost from the Persian Gulf to country i, and F_j^i is the freight costs from country j to country i. In mid-1978, the principal reference price for oil as established by OPEC was $12.70 per barrel of Arabian light (34° American Petroleum Institute (API) gravity) crude available fob at Saudi Arabia's Quoin Island terminal in the Persian Gulf.[1] Map 5.1 illustrates the January 1978 worldscale flat rates (in dollars per barrel) for shipping oil from Quoin Island to major oil-importing ports. (Worldscale rates are established semiannually by the International Tanker Nominal Freight Scale Association and serve as the basic reference listing of tanker rates. The worldscale rate listed for each route is hypothetical in the sense that it is an approximation of the cost of shipping a barrel of crude oil on a standard size—19,500 long ton—tanker. Actual rates are expressed as a percentage of the worldscale rates. This percentage varies, depending on supply and demand conditions in tanker markets.)

Because the U.S. ports are among the most distant from the Persian Gulf, the cost of shipping Persian Gulf oil to the United States (and Canada) is higher than the cost of shipping it to other major world markets. Conversely, because of Mexico's proximity, tanker costs from Mexico to the U.S. Gulf Coast are lower than to any other large oil-importing market. These two facts imply that as long as world oil prices are determined by the Persian Gulf price plus freight, Mexico will receive the highest fob price on the oil it sells to the United States.

FREIGHT COSTS

During the 1950s, the world tanker market had a relatively simple structure. At that time, the United States and Japan were importing

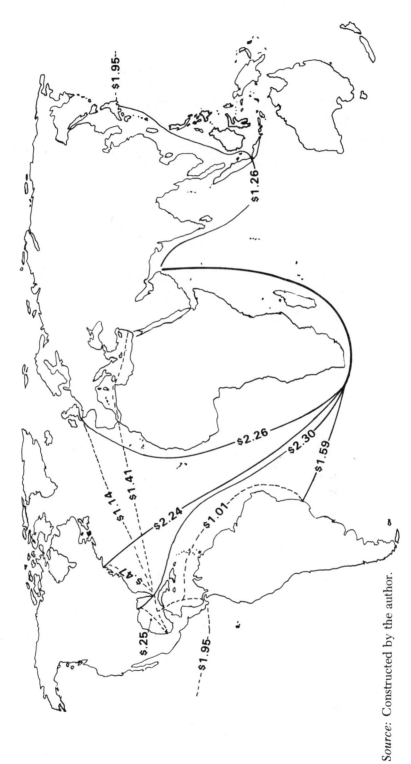

Map 5.1: Worldscale Flat Rates
January 1978 U.S. dollars per barrel

$1.95

$1.26

$2.26

$2.30

$1.59

$1.14

$1.41

$2.24

$1.01

$.47

$.25

$1.95

Source: Constructed by the author.

relatively little oil, and thus, the major route of the world oil trade was from the Persian Gulf to Western Europe via the Suez Canal. Due to Suez's draft limitations, the world's tankers were mostly in a single size category of between 20,000 and 40,000 deadweight tons.[2]

The situation began changing rapidly in the 1960s, as a result of the demonstration that supertankers enjoyed huge scale economies, the development of major North and West African oil export centers, the gradual (initially) but steady emergence of the United States as a major oil importer, and, after 1967, the prolonged closure of the Suez Canal. Together, these factors meant that over any given route, the freight rate would be influenced by the length of the contract (spot, consecutive voyage, or time charter), the size of the tanker, and the current supply-demand balance for oil and for oil tankers.

The prolific growth in the use of supertankers (200,000 deadweight tons and over—five to ten times larger than the typical tanker operating during the late 1950s) has been the most significant development in the world tanker market. With the sharply increased volume of oil traded internationally and with the greater distances involved in the trade patterns that emerged during the 1960s, the scale economies associated with supertankers enabled shippers to save large sums in the transport of oil. Although the trend towards supertankers was well underway before the 1967 closing of Suez, the shift to supertankers was accelerated by that crisis: oil importers discovered that the large tankers could bypass the Suez Canal, sail around Africa's Cape of Good Hope, and still deliver oil at low freight costs. Hence, while in 1967, the 210,000 deadweight ton *Idemitsu Maru* was the largest ship afloat, by 1969, there were some 35 tankers in the 200,000 to 326,000 deadweight ton range, and several hundred more were on order.[3]

As a result of the scale economies made possible by supertankers, the freight component in crude oil pricing has been considerably reduced. For example, the June 1978 worldscale rate for the shipment of oil from the Persian Gulf to Japan is $1.26 per barrel.[4] However, the actual rates paid by shippers are expressed in terms of a percentage of the worldscale rate for that route. Thus, in February 1978, a 50,000 deadweight ton tanker carrying crude from the Persian Gulf to Japan quoted a spot (one-trip charter) rate of worldscale 80 (WS 80). This means that the actual rate paid was 80 percent of the worldscale rate, or about $1.01 a barrel. In the same month, a 130,000 deadweight ton tanker chartered for the same route rented for WS 24, or $.30 per barrel, while a 225,000 deadweight ton tanker charged WS 18, or only $.23 per barrel.[5] Such enormous scale economies explain the rapidity of the shift to supertankers.

Although supertankers offer substantial economic benefits, their enormous size imposes some serious operational limitations. Larger tankers are wider, longer, and have deeper drafts; vessel draft represents the primary physical limitation on existing natural ports—although length, stopping distance, and turning restrictions are also important. The shortage of natural deepwater ports large enough to accommodate supertankers led, during the 1960s, to the construction of offshore deepwater terminals near major shipping centers throughout the world. With the important exception of the United States, most of the world's major oil export and import centers now have port facilities capable of off-loading ships of at least 150,000 deadweight tons.[6] Deepwater ports at the major import centers of Rotterdam, Tokyo, and Yokahama allow giant supertankers to unload their crude directly at the consuming area and thus reduce the overall costs of shipment. In sharp contrast, no U.S. East or Gulf Coast port can accommodate tankers over 80,000 deadweight tons.[7] Furthermore, the establishment of deepwater terminals in the United States continues to be delayed by a host of environmental and political concerns. As a result, most of the Persian Gulf and North and West African oil imported into the U.S. East and Gulf coasts must use transshipment terminals located in the Caribbean.

The transshipment terminal closest to U.S. Gulf Coast ports is in the Bahamas, where Burmah has spent $65 million to provide port facilities for tankers of up to 440,000 deadweight tons.[8] In the Netherlands Antilles, Shell and Exxon have established large transshipment facilities at Curaçao and Aruba, respectively. These two terminals and a third at Bonaire can handle ships of up to 500,000 deadweight tons.[9] At each of these "superterminals," oil is transferred from the larger carriers to smaller feeder vessels capable of unloading in U.S. ports. In short, the total transport costs from Nigeria, Libya, or the Persian Gulf to U.S. East or Gulf Coast ports is the sum of three items: (1) the cost of shipment from the exporting country to the transshipment terminal, (2) terminal charges for cargo transfer, and (3) the cost of transshipment from the terminal to U.S. ports. Table 5.1 illustrates the sharply higher freight costs incurred by U.S. oil importers because of the need to transship most non–Western hemispheric oil imports: transshipment raises transport costs roughly 35 percent on the Persian Gulf–U.S. Gulf of Mexico route; the percentage cost rise is considerably higher on oil shipped to the United States over the much shorter routes from North and West Africa.

The use of supertankers has led to a worldwide reduction in oil transport costs. However, the inability of U.S. East and Gulf Coast ports to unload supertankers means that transport costs to the United States are substantially higher than to Europe, Japan, or even to eastern Canada (see Table 5.2). The coupling of a Persian Gulf base price with higher freight charges means that the delivered price of imported oil is substantially higher in the United States than in other world markets (see Table 5.3). This fact should add to the economic attractiveness of the U.S. market for Mexico.

TABLE 5.1

Cost of Shipping Crude Oil from the Persian Gulf to the U.S. Gulf of Mexico in Early 1978
(per barrel)

Item	Cost
Persian Gulf to Caribbean terminal	$.59
Fees at transshipment terminal	.10
Transshipment terminal to U.S. Gulf Coast	.22
Total	.91

Note: Entries are based on January 1978 worldscale values and assume a 250,000 deadweight ton tanker chartered at WS.25 from the Persian Gulf to the Caribbean and a 50,000 deadweight ton tanker chartered at WS 85 for the shuttle between the transshipment terminal and New Orleans.

Source: Compiled by the author based on interviews and data from *Worldwide Tanker Nominal Freight Scale "Worldscale"* (London: International Nominal Freight Scale Association, January 1, 1978).

MEXICO'S NET GAIN FROM EXPORTING ITS CRUDE OIL TO THE UNITED STATES

In order to take full advantage of its locational economies and the prevailing worldwide pattern of delivered crude oil prices, Mexico must ship its crude oil to U.S. Gulf of Mexico ports on a fleet of tankers designed to minimize transport costs. The relatively short delivery distances involved in the intra-Caribbean oil routes and the constraints imposed on tanker size by available U.S. ports preclude the use of large (greater than 80,000 deadweight ton) tankers.[*]

Table 5.4 presents calculations (using the formula $P_j^i = P_{PG} + F_{PG}^i - F_j^i$) of Mexico's fob revenues from the sale of Arabian light quality oil in a variety of world markets that are likely to be interested in purchasing large quantities of Mexican oil. These calculations confirm that Mexico will reap the largest per barrel revenues by exporting its oil on small tankers to U.S. Gulf of Mexico ports. Mexico is in the midst of a program to expand the size of its ports. However, its ports are presently unable to load tankers larger than 60,000 deadweight tons. Thus, until port expan-

---

[*]Venezuela, presently the largest Caribbean oil exporter, has long recognized the economic realities of the intra-Caribbean oil trade by limiting the size of the tankers it uses. Almost half of Venezuela's crude oil and refined products are shipped on vessels of between 20,000 and 50,000 deadweight tons; another 30 percent is shipped on vessels in the 50,000 to 80,000 deadweight ton range. (See "Caribbean Report," *Seatrade*, February 1978, p. 24).

sion is completed, the first column of entries in Table 5.4 is the relevant one for assessing the relative economic attractiveness to Mexico of different export markets. The entries in column 1 show that Mexico would net an additional $.71 to $2.16 per barrel by exporting its oil to the U.S. Gulf Coast ports. The net economic advantage to Mexico from exporting its oil to the U.S. Gulf Coast would be reduced once Mexican port expansion is completed. Nevertheless, the entries in column 2 of Table 5.4 demonstrate that these advantages remain significant—ranging from $.27 to $.85 per barrel.

Mexico's publicly announced plans project that roughly half of its oil exports will ultimately be sold to the United States, Canada, and South America, and the other half will be destined for Europe and Japan.[10] To accomplish this goal, Mexico has begun to acquire a fleet of tankers. With

TABLE 5.2

Freight Costs from the Persian Gulf to Major Oil-Refining Centers

Persian Gulf to	Worldscale Flat Rate	Actual Rate (approximate)
New Orleans	$2.30	$.91
Eastern Canada	2.30	.58
Rotterdam	2.26	.57
Yokohama	1.26	.32

Note: The worldscale rates are for a 250,000 deadweight tanker in January 1978.

Source: Compiled by the author based on data in *Worldwide Tanker Nominal Freight Scale "Worldscale"* (London: International Nominal Freight Scale Association, January 1, 1978).

TABLE 5.3

Delivered Price of Arabian Light Quality Crude
(Persian Gulf price plus freight)

Destination	Persian Gulf Price	Actual Freight Rate	Total
New Orleans	$12.70	$.91	$13.61
Eastern Canada	12.70	.58	13.28
Rotterdam	12.70	.57	13.27
Yokohama	12.70	.32	13.02

Source: Compiled by the author based on data in *Worldwide Tanker Nominal Freight Scale "Worldwide"* (London: International Nominal Freight Scale Association, January 1, 1978).

TABLE 5.4

Mexico's FOB Revenues on Arabian Light Quality Oil Exported to Major World Markets (net per barrel foregone revenues compared with exporting oil on small tankers to the United States

| | Mexico's FOB per Barrel Revenues (foregone net revenues) | |
| | Small Tankers (less than 80,000 | Large Tankers (250,000 |
Market	deadweight tons)	deadweight tons)
U.S. Gulf of Mexico (New Orleans)	$13.38	$13.23 (-.15)
Eastern Canada (Nova Scotia)	12.67 (-.71)	13.11 (-.27)
Brazil (Salvador)	12.17 (-.21)	12.85 (-.53)
Western Europe (Rotterdam)	12.22 (-.16)	12.98 (-.40)
Japan (Yokohama)	11.22 (-.16)	12.53 (-.80)

Notes: (a) Calculation of fob per barrel revenues is based on the formula

$$P_j^i = P_{PG} + F_{PG}^i - F_j^i$$

The formula assumes WS 92 for small tankers and WS 25 for large tankers. Worldscale rates are for January 1978; (b) foregone net revenues is calculated by subtracting Mexico's fob per barrel revenues from sales by small tankers to the U.S. Gulf of Mexico from fob per barrel revenues in all other cases.

Source: Compiled by the author based on data in *Worldwide Tanker* "Worldscale" (London: International Nominal Freight Scale Association, January 1, 1978).

the aid of Italy's government-controlled Italcantieri shipyards, Mexico's largest shipbuilder, Astilleros de Vera Cruz, is currently building a series of 43,700 deadweight ton tankers for Pemex.[11] Mexico has also arranged a "ships for oil" barter deal with Brazil—Brazil will supply eight 17,900 deadweight ton tankers in exchange for $100 million worth of Mexican oil.[12] The entries in Table 5.4 confirm that it is very uneconomical to use small tankers on long transoceanic voyages to Europe or Japan. The fact that Pemex's present emphasis is on acquiring a fleet of small tankers suggests that its profit-oriented managers may, in fact, be preparing to place greater emphasis on encouraging more lucrative oil exports to the United States.

NOTES

1. "Selected Crude Oil Prices," *Petroleum Economist*, June 1978, p. 273.

2. Louis Bragaw, *The Challenge of Deepwater Terminals* (Lexington, Mass.: D. C. Heath, 1975), p. 28.

3. Taki Rifai, *The Pricing of Crude Oil* (New York: Praeger, 1975).

4. *Worldwide Tanker Nominal Freight Scale "Worldscale"* (London: International Nominal Freight Scale Association, January 1, 1978).

5. *Shipping Statistics and Economics, February 1978* (London: H. P. Drewry, March 1978), pp. III–IX.

6. Toby Winters, *Deepwater Ports in the U.S.* (New York: Praeger, 1975), p. 55.

7. Ibid., p. 49.

8. "Burmah's Bahamas Terminal," *Seatrade*, January 1978, p. 24.

9. "Caribbean Report," *Seatrade*, February 1978, p. 139.

10. "Down Mexico Way," *Seatrade*, December 1977, p. 7.

11. "Builders and Repairers Look to New Facilities," *Seatrade*, February 1978, p. 123.

12. "Brazil," *Seatrade*, January 1978, p. 81.

6

MEXICO'S PROBLEMS
AND POTENTIAL

Mexico is a poor country with a rapidly growing population of 64 million (as of 1978), extensive unemployment and underemployment, a distribution of income and wealth that is among the most skewed (in favor of the wealthy) in the world, a large native Indian population that has yet to be assimilated, and an inefficient, nearly feudal agricultural sector that employs nearly half of the labor force but which produces only 10 percent of the GNP and receives a mere 6 percent of national income. The most important policy question facing Mexico today is whether the speedy development of its enormous petroleum reserves offers a means for alleviating these serious socioeconomic problems and advancing Mexico's development as a successful, modern nation? In particular, can Mexico use its oil wealth to finance the capital investments necessary to transfer large numbers of underemployed workers from the agricultural sector to much more productive jobs in the industrial and service sectors? Since assuming office in December 1976, the administration of President Portillo has premised its policies on the assumption that the answers to these questions are affirmative. Nevertheless, there are many skeptics—both in Mexico and in the United States—who doubt that petroleum will make an appreciable positive contribution to Mexico's development. The following excerpts from a *Wall Street Journal* article, entitled "Despite Rising Wealth in Oil, Mexico Battles Intractable Problems," summarizes the pessimistic outlook:

> It is becoming clear, in the words of a U.S. State Department economist, that "the problems of Mexico won't be solved by oil, and its major

---

The introduction to this chapter was co-authored by Thomas Sadler; Stephen Davis co-authored the section entitled Mexican Politics and Mexico's Petroleum. Catherine Rau and Jesús Velasco provided useful research for the entire chapter.

problems may well be worsened by it." Indeed, adds a Commerce Department analyst in Washington, "There is a growing school of thought that oil could be very harmful to Mexico right now. It's providing an excuse—a crutch—not to solve the basic structural problems." Some even compare Mexico's breakneck expansion of oil sales to a derelict selling his blood in the morning to buy an afternoon bottle of wine. . . .

"There is no way that oil—its extraction, its refinement or its marketing—will improve employment in Mexico," says the skeptical State Department economist, who adds, "Given the way Mexico is organized, its (petroleum) will worsen income distribution. . . .

Already the tenth largest in the world, Mexico's population is expected to double by the year 2000. This surge—due to a high birth rate coupled with a relatively low death rate—is probably Mexico's worst problem and the one least solvable by oil money. . . .

While oil revenues apparently can do little to solve such problems, they clearly can do much to aggravate some others; corruption, inflation and, subsequently, the distribution of income. The large influx of money understandably arouses widespread suspicion in a country that, ever since 1950, has seen strong economic development accompanied by deterioration in income distribution and overt corruption. . . . As a popular expression puts it, "Oil, a gift of the gods, is the temptation of the Devil."

To the extent that oil money isn't used to service foreign debt, the funds also could become a temptation to spend—and thus a fuel for inflation. . . .

In the face of all these mounting pressures, Mexico's plan for stability has long been clear, most observers say. And it isn't oil.

"It's the same way the Irish and the English and the Germans and the Italians all solved their problems in the late 19th and early 20th Centuries, when their economies were transforming from agrarian to industrial," Prof. Blair of the University of Texas says, "Between 1890 and 1920 they sent 20 million immigrants to the U.S."[1]

Given the array of socioeconomic problems summarized by the *Wall Street Journal*, one would be foolhardy to state with certainty that rapid development of Mexico's petroleum wealth guarantees the nation greater stability and prosperity in the years ahead. Nevertheless, there are several reasons for suspecting that the *Wall Street Journal's* analysis overstates Mexico's problems and downplays its basic strengths. First, while there are no doubt many instances where some of those in Pemex or in the Mexican government have been either incompetent or corrupt, the same sort of observation would be true for most large organizations or governments. The real issue is whether the situation in Mexico is much worse than the norm elsewhere. Unfortunately, there is no objective comparative evidence. The comments of many U.S. academics and journalists (and by many Mexicans trained in the United States) seem to suggest that the problems of official incompetence and greed are especially difficult in Mexico. However, interviews with more than ten of Pemex's U.S. subcontractors uniformly yielded comments that Pemex managers are superior and more honest than the personnel of all other national oil companies with which they have worked.* The subcontractors' evaluations are also buttressed by key energy policy makers in other Latin American countries, especially Venezuela, who regard Pemex as the model their national oil companies should try to emulate.

Second, the coincidence of two events—Portillo's succession to the presidency and the beginnings of a rapid expansion in southeastern Mexico's petroleum production—have already precipitated a dramatic reversal in Mexico's worldwide economic status. Mexico was nearly insolvent at the end of the presidency of Echeverría: On August 31, 1976, the government was forced to allow the exchange rate for the peso, fixed at 12.50 per U.S. dollar since 1954, to float. Within two months, the price of a dollar exceeded 25 pesos, and Mexico was having great difficulty refinancing its $26 billion foreign debt. Because of economic and financial uncertainty, new foreign investment was unavailable, and domestic investors attempted to protect their fast-eroding assets by investing them in foreign markets.

Since 1977, the effective leadership by the fiscally prudent Portillo, coupled with modest oil exports (valued at around $1 billion in 1977 and $2 billion in 1978), has put an end to the flight of domestic capital and encouraged a modest resumption of foreign investment. However (assuming maintenance of 1978 world oil prices), if Mexico chooses to continue to push development for its petroleum industry at the 1978 rate, by 1985, it should be grossing $15 billion on oil and natural gas exports of at least 3 million barrels per day (MBD); by 1990, revenues should hit $28 billion on oil exports of 5.5 MBD. Because they would eliminate all fears of Mexico's insolvency, revenues of this magnitude would assure that if Mexico wants massive new private foreign investment, Mexico will get it.

The greatest single cause of concern for Mexico's leaders has been its soaring rate of population growth. Mexico's population of 64 million has been growing at an annual rate of 3.2 to 3.5 percent in recent years. Even if one believes the government can achieve its proclaimed goal of reducing the rate of population to 1 percent by the end of this century, Mexico's population would still be over 100 million. Less optimistic projections, probably more plausible, are that Mexico's population will be

---

*Three of these interviews were conducted by the author on the scene at Reforma; the remainder were at the subcontractors' home offices and were conducted under the author's superision as part of a project sponsored by an agency of the Venezuelan government.

between 120 million and 130 million by the dawn of the twenty-first century.[2]

The problems of unemployment and underemployment are closely related to the problems of population growth. Of Mexico's 17.5 million work force in 1977, an estimated 8.5 million were either unemployed or underemployed. Because roughly half of Mexico's present population is under 15 years old, the demand for jobs will continue to soar for another 15 years or more regardless of the success of population control policies. An estimated 800,000 young Mexicans will enter the job market in 1978.[3] Assuming the Mexican economy returns to its traditional 6 percent growth path, annual job creation would be only about 350,000. In consequence, unemployment and underemployment definitely seem likely to grow in the near term.

Growing Mexican unemployment creates obvious problems for the United States. There are already an estimated 6 million to 10 million Mexicans illegally residing in the United States, and in 1977, the U.S. Border Patrol caught and turned back 750,000 illegal aliens.[4] Within Mexico, the greatest poverty and unemployment is found in rural areas. Thus, there has been a massive migration toward the more prosperous urban areas, principally Mexico City, Guadalajara, and Monterrey. Arriving without money or jobs, the peasants have packed themselves into vast slum areas, like the "lost cities" that encircle Mexico City. According to the *Wall Street Journal:* "One of the worst, Nezahualcoqotl, has over a million residents, many of them living in squalid shacks fashioned from cardboard pieces of billboard and old car parts."[5] The outlook for the new urban poor is bleak. A recent U.S. population projection states that within 25 years, Mexico City will be the largest city in the world, bulging with a population in excess of 31 million.[6] Already, Mexico City is known as the world's largest *poor* city. According to Alan Riding, Mexico City displays all of the symptoms of poverty: "a population growth rate of 3.2 percent per year despite a new birth control program, malnutrition affecting 60 percent of children under five, inadequate housing for 75 percent of the population, a functional illiteracy rate of 60 percent for adults, and a high infant mortality rate due to poor hygiene."[7]

Throughout the 1960s and 1970s, large (relative to the arable land mass) and/or rapidly growing populations have been viewed as a cancer, making it nearly impossible for a country to escape the confines of poverty. However, the rather recent successes of Japan, Taiwan, South Korea, and Singapore—as well as the more established successes of several Western European nations and the United States—suggest that this view may be too simplistic. The real shortcomings with this "population bomb" analysis is its asymmetric view of the human animal: humans are viewed almost exclusively as parasitic consumers, whereas, in fact, they can also be productive producers. In short, nations do not suffer poverty because they have too large a population but because they are unable to find productive employment for sufficient numbers of their people. In view of Mexico's enormous petroleum wealth, the key policy issue will be whether it chooses to use its oil revenues to follow the course outlined by Gustavo Romero Kolbeck, president of the Banco de Mexico:

> The Mexican economy is a well-diversified one, in regard to internal production as well as in relation to the structure of our exports. For many years Mexico has been able to develop its various sectors of the economy, such as agriculture, steel, petrochemical industries, tourism, manufacturers and others, and it is our intention to continue to do so and not to base our future on any one single product. In the future, oil will play a major role and will provide Mexico with additional resources that can be invested not only in the field of energy but in many other sectors that will provide employment. We are determined not to allow the growth of any one single sector or product to distort the rest of the economy and possibly hinder our long-term development. To that end, the foreign exchange derived from oil exports will be channelled to productive investment in fishing, agriculture, industrial activities and other sectors, thus avoiding the export of capital from Mexico, which is still a developing country and in need of these funds.[8]

The skeptics have usually been correct in predicting that specific developing countries will not complete the metamorphosis into a modern nation. Nevertheless, when encouraged by effective political leadership, successful transformations have occurred. Judged by objective measures, such as per capita GNP ($1,120 in 1977) and literacy rates (84 percent in 1970), Mexico is already a fairly advanced developing country. Moreover, in addition to huge petroleum reserves, Mexico enjoys several advantages not present in other developing countries: it has a relatively advanced and diversified industrial sector (see Table 6.1), a fortuitous location next to one of the world's largest markets, and perhaps most important, her educated elites are bound together by a sense of their nation's great cultural and historical heritage. Thus, there is some prospect that a large influx of oil revenues will find an especially fertile climate in Mexico. Should this happen, Mexico's burgeoning population could prove an asset, assisting it to become one of the world's most influential nations.

THE CRISIS OF 1976 AND ITS AFTERMATH

Until the early 1970s, Mexico was widely regarded as a textbook example of a slowly but steadily modernizing developing country. The peso was stable; there had been no banking failures since 1937; the rate of

TABLE 6.1

Mexico's Manufacturing Production, 1972 and 1976

Products	1972	1976
Steel[a]	4,384	5,225
Copper[b]	59,588	82,915
Aluminum[b]	39,484	42,357
Grey cement[a]	8,593	12,578
Sheet glass[b]	86,765	126,434
Sulphuric acid[a]	1,518	1,953
Fertilizer[a]	2,232	2,537
Synthetic fiber[b]	122,071	197,016
Automobiles[c]	168,552	227,803
Trucks[c]	58,666	90,660
Refrigerators[d]	281	501
Television sets[d]	436	729

[a] Thousand tons.
[b] Tons.
[c] Units.
[d] Thousand units.
Source: "The Future for Mexico," *Euromoney* (supp.), April 1978, p. 26.

real economic growth had averaged nearly 6 percent annually for nearly 40 years; and per capita GNP was $1,200 when Echeverría assumed the presidency in 1970. For these reasons, Mexico's international financial reputation was good, its credit rating in international money markets was solid, and its access to external sources of credit was among the very best in Latin America. This favorable state of affairs began to deteriorate rapidly in late 1974 due to the confluence of two events: President Echeverría's too-ambitious program of social and economic reforms and worldwide recession, accompanied by plummeting commodity sales and prices as a result of the 1973–74 OAPEC oil embargo.

The basic concerns of President Echeverría's administration were to reduce Mexico's deep-seated problems of unemployment and the unequal distribution of income. To attack these problems effectively, Mexico needed to maintain rapid economic growth. But saddled with an austere federal budget during his first full year in office (1971), the new president presided over a sharp fall in the rate of economic growth. The growth of Mexico's GDP fell from a healthy 7 percent in 1970 to only 3.4 percent in 1971, a rate hardly higher than the rate of population growth.[9] Blaming the Mexican economy's poor performance on the alleged failings of the business community, Echeverría chose to reverse the economic decline by sharply raising federal spending. A large portion of the higher federal spending was aimed directly at aiding Mexico's disadvantaged groups. Initially, the strategy of massive new federal spending appeared to be successful: the growth rate of Mexico's GDP rebounded to 7.3 percent in 1972 and 7.6 percent in 1973. However, with the benefit of hindsight (and perhaps some appreciation of the "no such thing as a free lunch principle") it appears that Mexico's prosperity during 1972–73 was due to the worldwide commodity boom far more than it was due to President Echeverría's well-intended policies.

At the start of 1974, Mexico's economy appeared healthy—being only a modest oil importer, Mexico seemed to be largely immune from the soaring oil prices that had begun to wreak havoc in most of the world's major economies. But by the autumn of 1974, the Organisation of Economic Cooperation and Development (OECD) countries were in recession, and commodity prices and sales were plummeting. Ignoring the worldwide recession, the Mexican government continued to spend at record high levels. Since the value of Mexico's exports was falling sharply, there was a substantial worsening of Mexico's balance of trade, forcing the government's foreign borrowings to jump from $6.5 billion in 1973 to more than $10 billion in 1974. By 1976, the Mexican government's foreign debt had soared above $20 billion, inflation had risen to a 60 percent annual rate, and sensing that economic collapse was imminent, billions of Mexico's private capital was being reinvested overseas.[10] Richard Fagen has described how this crisis culminated:

> On August 31, 1976, in a move that caught many by surprise . . . the peso was devalued for the first time in 22 years. With the peso floating "like a stone" . . . multiple reactions and even panic ensued. While Mexican and foreign dailies headlined "turmoil," "hysteria," and "crisis," as much as four billion dollars fled the country seeking safe harbor in Texas banks and elsewhere. Investment slowed down, inflation accelerated, unemployment rose, and the whole complex set of mechanisms by which devaluation and resultant dislocations and hardships are passed disproportionately on to the poorer sectors of society came into play. Twelve days before leaving office, when Echeverría expropriated tens of thousands of acres of prime land in the northern state of Sonora and turned them over as small parcels to peasants, talk of a military coup was heard for the first time in recent memory.[11]

The presidency of Portillo began in the midst of this maelstrom in late 1976. Upon entering office, the new president had only two fragile tools with which to halt Mexico's crisis of 1976. First, he repeatedly emphasized his commitment to unite the nation behind a government that would practice fiscal prudence. In a manner reminiscent of Franklin D. Roosevelt, Portillo told the Mexican Legislature (September 1, 1978) in his first State of the Nation report:

Thirty-nine weeks ago . . . (t)he economy wás undergoing the most astute crisis this generation has ever known. Mexico, full of self-doubt, had reached the darkest moment of its crisis. . . . I did not promise miracles, knowing that in economic matters it is impossible . . . I offered a rational project of shared effort to handle the crisis. . . . Like all beginnings, this one is full of ambiguities, uncertainties and anxiety. But it also represents commencement and promise.[12]

Second, he held out the hope that Mexico's apparently large new oil reserves—their true scope was not yet recognized—might support substantial oil exports and thereby permit Mexico to service its large foreign debt.

Given the sudden exodus of investment funds in late 1976, it was inevitable that the Mexican economy would stagnate during 1977 and early 1978. Nevertheless, President Portillo's eloquent promises and hints of big oil immediately struck a responsive chord with the international bankers, who had so many billions precariously invested in Mexico. Their confidence grew when, as the months passed, the existence of huge oil reserves was confirmed and President Portillo showed that he did indeed have the power to cut Mexico's balance of trade deficit in half, to enforce the IMF's tight $3 billion limit on international borrowing, to stabilize the exchange rate, and to substantially reduce the government's share of GNP. Mexico's wealthy citizens began repatriating their foreign assets, and by early 1978, the world's major international banks were eager to advance new credits. By mid-1978, even though federal spending had been cut back substantially, the economic recovery was well underway. Mexico's real growth should exceed 5 percent in 1978, and much more rapid growth seems likely in 1979.[13]

MEXICAN POLITICS AND MEXICO'S PETROLEUM

Mexico has been governed by members of the Partido Revolucionario Institucional (PRI) for 40 years. The PRI is an unwieldy coalition of interests. Under the umbrella of nationalist ideology, three formal groups or sectors—peasant, labor, and middle (the word *class* is shunned)—vie for influence. The private sector, because it is officially regarded as nonrevolutionary, is not represented in the PRI's formal hierarchy. Nevertheless, historically, it has wielded a large share of influence over PRI policy.

The peasant's official party organization, Confederación Nacional Campesina (CNC), is probably the least effective lobby in the party. Because of agriculture's low productivity, CNC's members are generally without financial resources and, therefore, depend on the government for land, capital, equipment, and political support. As a result, Mexico's peasants have little bargaining power besides their ballots, and the PRI can usually take their ballots for granted, because Mexican farmers regard a vote for the "revolutionary" party as a patriotic act. "The poor," asserts the *Economist*, "are in the pocket of the PRI."[14] CNC leaders are chosen by the PRI/government elite and often come to their posts with little rural background.

The peasants have little real power within the PRI. However, since the revolution is popularly believed to have been undertaken by and for Mexico's poor, the peasants do command strong ideological support. It is, therefore, of more than passing interest when a peasant representative publicly challenges the PRI leadership's decision to promote petroleum exports. In October 1977, one such representative in the Chamber of Deputies, Victor Manzanilla Schaffer, declared that government policy should be "not to sell crude, rather to use it in our own country."[15] Schaffer also voiced opposition to a constitutional amendment ceding to Pemex broad land expropriation powers to facilitate petroleum exploration and development. The incident serves to demonstrate the scope of peasant interest in petroleum: farmers are nationalistic, concerned with access to inexpensive chemical fertilizer, as well as oil and gas income, and jealous of land tenure security. Reflecting the CNC's minimal influence within the PRI, the government's response to peasant concerns has been largely rhetorical. The Pemex-backed constitutional amendment was passed, but President Portillo did offer promises that "foreign exchange derived from oil exports will be channeled to productive investment in . . . agriculture . . . and other sectors."[16]

The Confederación Nacional de Organizaciones Populares (CNOP) is the PRI organization representing the 50-plus subgroups that constitute Mexico's middle sector. Because it is a heterogeneous lobby, the CNOP is often unable to reach a consensus, much less to mobilize its forces in defense of middle sector interests. Despite the fact that CNOP's bargaining resources are much more extensive than those of the peasantry, the PRI is usually capable of manipulating their use. For example, the largest component of CNOP is the union representing government employees, whose membership is dependent upon the regime for their jobs. More generally, the fact that the PRI controls licensing of many middle sector occupations makes CNOP subservient to the ruling party. To date, the middle sector has contributed quietly to the petroleum debate only to the extent of encouraging the flow of as much oil and gas income as possible into broad swaths of the Mexican economy. It cannot be expected to contest Pemex-backed policies in the future. However, as one might expect, it will work behind the scenes to gain the PRI's commitment to distributing more of the benefits from petroleum sales to its constituents.

Labor and business interests will have the greatest impact on the PRI's petroleum policies. The labor movement as a whole is represented in the PRI by the Congreso de Trabajo, established in 1966 as a united front for all labor groups. Its leaders are appointed subject to PRI approval, and in its formative years, dissidents were sometimes jailed. Despite an overall ideological consensus within the Congreso, cooperation between members is minimal, and the PRI leadership frequently plays unions against each other to maintain control.[17] The situation, however, is somewhat different in the petroleum industry.

The petroleum workers' union, STPRM, benefits from the popular memory of labor's assumption of control over the oil fields following President Cárdenas's 1938 nationalization decree. Pemex's *petroleros* are given a special place in Mexico's labor movement, and they have been singularly successful in maintaining relative unity and an ability to demand and receive highly favorable labor contracts. STPRM's members receive the highest wages and best fringe benefits in the entire Mexican labor sector.

Pemex's petroleros were, until recently, led by men of the "generation of 1938," whose guiding principle was resource nationalism. These were the old guard, who opposed exporting Mexico's petroleum, especially to the United States. STPRM's leadership has recently changed hands. The present union leadership—recognizing that the economic welfare of both members and leaders coincides with an expanded role for Pemex—appears willing to endorse Pemex's plans for rapid development of a substantial petroleum export capability. In return for its consent, STPRM's primary conditions are that its members be guaranteed continued hikes in relative wages and benefits, coupled with assurances of no layoffs. Because of Pemex's rapid growth, meeting these demands may not interfere too seriously with Pemex's efforts to improve labor productivity.

Ironically, it has been the "nonrevolutionary" private sector that, historically, has had the most influence on the PRI. Though officially excluded from the PRI's formal structure, there are at least three structural reasons why industry interacts closely with government. First, railways, communications, utilities, and several key industries, such as petroleum, are all in the public sector. Second, government pricing, credit, tariff, and investment policies have a direct impact on private business. Third, and probably most important, the PRI is able to control labor and thereby wins the gratitude of the private sector.

Of course, the relationship between private business and Mexico's government is reciprocal. Business helps to mobilize employment and capital—both of which are scarce in Mexico. Generally, business interests are enthused about, and willing to contribute toward, Pemex's plans to develop a large petroleum export capability. However, since the revenues from petroleum exports will flow initially into a public company, Pemex, Mexican business is exercising pressure to insure that there is no relaxation in present laws mandating extensive use of Mexican subcontractors. Business is also pressing for policies to assure that most of Pemex's oil profits will be recycled to stimulate broad-based development of the economy.

In sum, the PRI dominates Mexican politics, and the groups with the most power over the direction of the PRI favor vigorous development of Mexico's oil reserves, at least until Mexico reconsiders its national energy policy in 1980. Furthermore, since Mexico actually will be receiving and spending large oil revenues when the direction of its energy policy is reconsidered in 1980, it seems unlikely that the PRI's commitment to developing Mexico's petroleum potential will diminish. Nevertheless, because President Portillo has pushed through electoral reforms designed to strengthen the political opposition and transform the Chamber of Deputies from a rubber stamp "relic of the democratic ideal" into a "loud, critical voice that will act as a watchdog over the government," brief examination of the oppositions' energy views is merited.[18] Although these views are unlikely to have much importance in the near term, they could have an impact on the long-term evolution of Mexico's energy policy.

The oil issue has recently become the spearhead of leftist criticism against the government. In March 1978, an ad hoc coalition of 20 groups, constituting the National Front for the Protection of Natural Resources, staged a 5,000-person demonstration in Mexico City. Castillo, head of the Mexican Workers Party, used this occasion to denounce the "massive and indiscriminate" exploitation of petroleum by Pemex and to attack Mexico's subservience to the United States.[19]

Since Mexicans are ferociously nationalistic about their oil, it follows that the opposition would attempt to win support by appealing to resource nationalism. Because oil is the one issue that attracts a consensus, it has catalyzed Mexico's left into attempting a fragile union of their normally warring factions. Late in 1977, three leftist parties held a joint assembly to build the foundation of a united Partido de la Clase Obrera (PCO), based on Marxist-Leninist ideology. Though representing a significant concentration of effort, the new PCO still excluded more groups than it embraced. Nevertheless, all leftist parties have declared their opposition to the sale of oil and gas to the United States.

Castillo's Mexican Workers Party has been at the forefront of the petroleum opposition. Castillo denies Pemex's claim that it must either sell Mexican gas to the United States or flare it. Castillo's solution would be to shut in the wells and produce only to the level of domestic demand. According to Castillo, proposals to construct a natural gas pipeline from the Reforma fields to the United States and to bring more crude onstream for export will only serve to alleviate the U.S. energy problem and drain Mexico of petroleum. Moreover, Castillo asserts that Pemex's drive to produce more petroleum will create only minimal employment in the context of an enclave economy, where oil wealth remains segregated from the majority of Mexicans. The result could be a single-product economy, consisting of two separate societies: "one employed and benefitting from economic growth, and the other underdeveloped and relying on government handouts for survival."[20]

Another concern of the Mexican left is that Mexico, by being an energy lifeline for the United States, will, inevitably, be included in a U.S. security zone. Critics have gone so far as to compare the proposed natural gas pipeline to the United States with the Panama Canal, charging that a slice of Mexico would be mortgaged and its sovereignty sold to Washington.[21] President Portillo felt compelled to address this charge directly in October 1977, when he asserted that the "U.S. is not going to change its position of respect for the structure of peace and invade us. That would be going too far."[22]

The Mexican right has not embraced the petroleum issue to the same extent as the left. Electorally stronger than the groups on the opposite extreme, the rightist Partido Acción Nacional (PAN) has been split since 1976, when old guard conservatives balked at younger members' attempts to remake the party on a Christian Democrat model. Though PAN took four seats away from the PRI in the 1973 congressional elections, the virtual internal civil war kept the party from participating in the 1976 presidential race.

PAN is the second largest and second oldest party in Mexico. It has been termed the "revolutionary right" by some and a "confused party" by others.[23] It subscribes to a blend of Catholic, free enterprise, and nationalistic values, some of which may be in conflict when addressing the petroleum question. Assuming it reestablishes a workable measure of party unity, PAN may be expected to try to tap populist sentiment and pressure the government into guaranteeing a greater role for Mexican firms in petroleum development.

Mexico's electoral reforms will enlarge the Chamber of Deputies from 250 to 400 members; 300 of these members will be elected from separate constituencies on a winner-take-all basis, while the remaining 100 will be elected from five regional districts, according to proportional representation. At least half of the 100 regional seats are reserved for the PRI's opposition. Electoral reform seems certain to lead to a proliferation of parties and may lead to an upgrading of the Chamber of Deputies' importance. However, the PRI will remain firmly in control, since the electoral regulations provide that any party winning more than 90 seats under the winner-take-all system must automatically forfeit half of its deputies elected by means of proportional representation. The upshot will be that in the forseeable future, the PRI will be able to count on a significant majority of the Chamber's 400 seats and, as President Portillo has said, can "prevent the proportional principle, essentially a just one, from being converted into instability.[24]

There is reason, in fact, to believe that Portillo sponsored political reforms for the very purpose of revitalizing the PRI. After nearly 40 years without real competition, the PRI today is powerful but fossilized. Its role as a vehicle for political mobilization now comes into play only during elections; otherwise, the organization is barely distinct from government. Though PRI unresponsiveness at the grass roots has generally produced an attitude of apathy toward the political system, some Mexicans have found it to be a source of deep frustration. This frustration may, in part, explain the reappearance of small guerrilla groups and a growing opposition. An emotional issue, such as petronationalism, could, in these circumstances, ignite a significant challenge to PRI rule.

A rapid influx of oil wealth could also precipitate a period of serious political instability in Mexico. If economic growth is perceived as inequitable or if the majority of Mexicans experience only an insignificant improvement in their living standards, the government could become the target of widespread protest, possibly culminating in violence. President Portillo may be gambling on the belief that liberalizing the PRI and fostering a stronger opposition may strengthen both his party and his government. As Mexico watchers John and Susan Purcell have observed, "As long as the strength of the opposition parties is kept at tolerably low levels (that is, high enough to spur the PRI to greater activity but low enough so that PRI control is not seriously threatened), the PRI may gain by having its organizational capabilities reinforced."[25]

If the 1978 electoral reforms are something less than an invitation to political pluralism, they are, nevertheless, indicative of the way in which the PRI relies on cooptation to deal with dissent. Cooptation is the PRI's guiding strategy. The PRI has become expert in spotting dissent early and defusing it. As if following classical Clausewitzian tactics, the PRI drains militancy from the opposition by aiming at the movement's center of gravity—a corp of leaders in some cases, a principal grievance in others.

Given the PRI's past behavior and the limited nature of President Portillo's reforms, it is unlikely that the opposition, either right or left, will either dislodge the PRI from power or exert much influence on the petroleum debate. In the absence of further structural changes, the PRI will continue to have a solid majority in the Chamber of Deputies and to name Mexico's president. It is conceivable that in the new, more open political environment, the PRI will find itself under more pressure to acknowledge opposition criticism. But if previous behavior provides any guidelines, it is unlikely to modify actual policies in more than superficial or rhetorical fashion.

The PRI is more than the sum of its parts. Its ideology of nationalism is derived from all its constituent groups, as is its concern with economic development and social equity. But it is the party's juggling of interests that makes it distinctive. The politically weak and potentially dissident lobby groups are incapable of seriously influencing the PRI's petroleum policies, while the more powerful groups are either divided, coopted, or supportive of Pemex. Thus, as long as it is capable of satisfying the minimum demands of labor and business, the PRI faces little challenge to present plans to push petroleum development from within the party. In addition, the PRI appears readily capable of fending off dissent from other parties, merely by maintaining its strong rhetorical commitment to resource nationalism.

SPECIAL CONCERNS

In less than two years, the combination of President Portillo's fiscal conservatism and his skillful and persistent emphasis on Mexico's rapidly growing petroleum export potential has caused a favorable transformation in Mexico's economic outlook. Fueled by expanding petroleum export revenues, the Mexican economy is entering the first stages of an era of rapid real growth. Moreover, expanded foreign exchange earnings will allow the nation to lessen its foreign debts and thereby gain greater independence from the sanctions of international bankers. Yet, somewhat paradoxically, many members of Mexico's educated elite, who traditionally have lamented their country's subservience to foreign businesses or countries, argue that their president is moving too fast. In particular, while not attacking Pemex's plans to expand crude oil output to 2.2 million barrels per day by 1980, they maintain that a go-slow approach is needed in the post-1980 period. There appear to be several motives behind this reluctance to have Mexico seize the opportunities now available to it.

First, many college-educated (thus, by definition, privileged) Mexicans appear to have a schizoid view of economic growth. On the one hand, they assert that growth is necessary to alleviate the severe poverty most Mexicans suffer. But on the other hand, they fear that too rapid growth could disrupt the social compact that has kept Mexicans at peace with one another for the past 50 years. In view of Mexico's turmoil during the last years of Echeverría's administration, it is natural for Mexicans to exhibit concern about their nation's social stability. However, since the existence of huge amounts of petroleum is already well known—and, therefore, the expectations of many Mexicans have already risen to high levels—there is reason to suspect it is already too late for Mexico to adopt a go-slow petroleum development strategy. To do so might precipitate the very instability such a strategy is supposed to prevent.

Anti-Americanism provides a second possible motive for advocating a go-slow approach to petroleum development. Large exports to the United States are inevitable if Mexico is to develop its petroleum potential as quickly as is possible and if it is to net the highest profits on each unit of crude oil and, especially, natural gas that it exports. But large imports of Mexican petroleum would help to alleviate several of the energy problems of the United States (see Chapter 7 for elaboration). Recognizing this, some Mexicans seem willing to deny their country the full advantage from petroleum sales in order to wield an oil weapon against the United States. This study is not the appropriate forum for asking whether there is justification for Mexican anti-Americanism. However, it would be prudent for those Mexicans advocating minimal (or no) petroleum exports to the United States to answer two questions directly. First, is it worthwhile for a poor, debt-ridden country like Mexico to sacrifice large foreign exchange earnings in order to aggravate a much more prosperous nation like the United States, whose power vis-a-vis Mexico derives in large part from Mexico's poverty and huge foreign debt? Second, suppose (what is unlikely) that by denying petroleum exports, Mexico could seriously weaken either the United States economy or its oil security. Is it in Mexico's interest to have a politically, militarily, or economically unstable major power as its neighbor?

A mixture of Mexico's long-standing commitment to resource nationalism and the fear that its oil reserves will be depleted quickly is a third motive for those opposing rapid expansion of Mexico's petroleum industry. President Portillo tried to reassure the nation on this point in a speech delivered on September 1, 1978. After announcing that Mexico's proved petroleum reserves had risen from 16 billion to 20 billion barrels in the preceding six months and that potential reserves had risen from 120 billion to 200 billion barrels, the President said: "We can assure you that Mexico has enough oil and gas to last us well into the twenty-first Century. . . . [Mexico's oil income] will be earmarked for already established national priorities. Others won't be invented just because these resources exist."[26]

Mexico projects that its combined production of crude oil and natural gas will total about 2.5 MBD in 1980. At this rate of production, Mexico's presently known potential reserves would last 220 years; these reserves would last 70 years if production was to triple to 7.5 MBD. Of course, since Mexico's oil boom is still in its early stages, there is a strong probability that the inventory of potential petroleum reserves will continue to grow rapidly as additional exploratory drilling progresses. In short, Mexico's likely petroleum reserves are already so large that resource exhaustion is an irrelevant consideration over any reasonable planning horizon.

A fourth concern that some Mexicans have about expanding petroleum production is that it will make Mexico more important to the security of the United States, and consequently, the United States will react by extending its control over Mexico. Without doubt, the United States would take steps to assure that there are no interruptions in the flow of petroleum exports if Mexico was to become an important source of supply. However, events since the OAPEC oil embargo strongly suggest that these measures would enhance rather than reduce Mexico's independence vis-a-vis the United States. More precisely, in the few short years since the OAPEC embargo, all large OPEC oil exporters, and preeminently the two largest, Saudi Arabia and Iran, have become recognized as important actors on the world stage. Having demonstrated their power to disrupt the world's economy by cutting back the flow of oil exports, the OPEC countries have forced the large industrial countries to discover that OPEC interests can no longer be treated with impunity. The new respect is not just symbolic but real. Thus, OPEC's principal members are now consulted regularly about how to remedy international financial problems, and Arab nations (still officially at war with Israel) discover that Washington responds favorably to requests for advanced military hardware. The huge asymmetry in the relative power of Mexico and the United States explains why relations between these two neighbors have always been one-sided—Mexico has always been negotiating from weakness. U.S. dependence on large imports of Mexican petroleum would, for the first time, cause some redress of the economic and power imbalance. The experiences of Saudi Arabia and Iran suggest that U.S. negotiators will be much more responsive to Mexico's demands if Mexico can trade something the United States wants and needs.

A fifth concern of those Mexicans who oppose the rapid expansion of petroleum production is that the Mexican economy will be unable to absorb the resultant revenues efficiently. Those who express this view argue that the result will be massive inflation and waste—they point to Venezuela as their chief example.

Following the onset of the OAPEC embargo, the price of Venezuelan oil exports increased dramatically. There was a threefold rise in Venezuela's oil prices just during the first quarter of 1974, and by April 1975, export prices had quintupled from 1972 levels.[27] Since Venezuela's oil exports averaged about 2.5 MBD in the early 1970s, soaring prices had a dramatic effect on its petroleum revenues—the value of Venezuela's oil exports rose from about $3 billion in 1973 to nearly $9.5 billion in 1974.[28] In a nation of only 11.5 million and with a nearly nonexistent industrial and manufacturing sector except for oil production and refining and iron ore mining, it is not surprising that this enormous, unanticipated windfall led to large waste and high inflation.

The situation in Mexico is very different from that in Venezuela at the time of the embargo. Mexico's 1978 population and labor force are five times as large as Venezuela's in 1973, and its industry is much more diversified. Unlike Venezuela, which was in the midst of a disruptive effort to nationalize its petroleum (and iron ore) industry in 1974, Mexico has been running its own oil industry for 40 years. Moreover, because Mexico's higher oil revenues will be due to rising output rather than unanticipated price hikes, Mexico has time to plan how to use them. Finally, and most important, even if Mexico chooses to expand its petroleum industry rapidly, Mexico's total and per capita oil revenues will grow far more slowly but more steadily than did Venezuela's. For example, if Mexico would choose to export an additional 500,000 barrels per day annually beginning in 1980, its oil revenues would increase (assuming maintenance of a $14 per barrel price) at an annual rate of $2.5 billion. This amounts to an annual increase of $36.50 per capita (or less than 3 percent of per capita income) in a nation of 70 million. In comparison, Venezuela's per capita oil revenues soared by almost $550 (or 42 percent of per capita national income) in 1974.[29]

In the aftermath of the OAPEC embargo, Mexico became the first country to discover and develop large, and prior to 1972, totally unsuspected petroleum reserves. As such, it is in a unique position to reap the benefits from rapidly growing petroleum exports—not being an OPEC member, it is free to export all the oil it can produce at the Persian Gulf price plus freight. Each of the five misgivings about Mexico's developing its petroleum reserves at a rapid pace appears to be either greatly exaggerated or misplaced. Thus, it becomes evident that the real source of the internal debate is disagreement as to whether Mexico should take the gamble and seize the opportunities for economic growth and modernization that access to really "big oil" makes possible. Recognizing that Mexico's present advantages could prove transient should large petroleum reserves be found in other countries, the Portillo government sees the necessity of taking the plunge.* In the words of Pemex's Director-General Serrano:

---

*Potentially enormous reserves of Canadian natural gas may offset the biggest near-term threat to Mexico's U.S. markets. John A. Masters, president of Canadian Hunter Exploration, Ltd., and the leading proponent of the hypothesis that Alberta has enormous deposits of relatively high cost—$2 per 1,000 cubic feet—natural gas has written:

Strong geological evidence points to the probable existence of an enormous gas accumulation trapped in the deepest part of the Alberta syncline and its extension into British Columbia.

This area, termed the Deep Basin, covers some 26,000 square miles. The gas interval includes almost the entire Mesozoic clastic section which reaches a maximum thickness of 15,000 ft. Reserve potential is so large as to alter significantly the energy supply estimates for North America. . . .

The entire Deep Basin Area is indicated to contain potential recoverable gas

The social cost of our failing to follow a dynamic production policy will be very great. Every industrial development program for primary and secondary petrochemicals and associated manufacturers will be slowed down in the short and medium term; not only will we be unable to catch up with the developed countries and fight our way into the respective markets, but we shall see other developing nations which are also oil producers and which are working in the same direction we are ... inevitably catch us up, thus making our position very difficult and blocking our ambition of becoming, for many, many years, a power in the petrochemicals world.[30]

Based on the considerable progress of President Portillo's first two years in office, there is reason to suspect that Mexico may succeed.

NOTES

1. John Huey, "Despite Rising Wealth in Oil, Mexico Battles Intractable Problems," *Wall Street Journal*, August 30, 1978, p. 1.

2. David Gordon, "Clouds of People," *Economist*, April 22, 1978, p. 7.

3. *Economist* Intelligence Unit, *Quarterly Economic Review of Mexico*, first quarter, 1978, p. 9.

4. Ibid., third quarter, 1977, p. 8.

5. John Huey, "Mexico's Economic Ills Could Topple Coalition If Workers,, Poor Rebel," *Wall Street Journal*, August 8, 1977, p. 1.

6. Ibid.

7. Alan Riding, "Silent Invasion: Why Mexico Is an American Problem," *Saturday Review*, July 8, 1978, p. 15.

8. "The Future for Mexico," *Euromoney* (supp.), April 1978, p. 6.

9. David Gordon, "Mexico, a Survey," *Economist*, April 22, 1978, p. 16.

10. "The Future for Mexico," op. cit. p. 2.

11. Richard R. Fagen, "The Realities of U.S.-Mexican Relations," *Foreign Affairs*, July 1977, p. 694.

12. President Jose Lopez Porfillo, *First State of the Nation Report*, (September 1, 1977) reprinted in *Comercio Exterior de Mexico*, September 1977, pp. 330, 335.

13. "Mexico's Oil, Foreign-Exchange Reserves Soar; President Cites Economic Recovery," *Wall Street Journal*, September 8, 1978, p. 6.

14. Gordon, op. cit., p. 11.

15. "Pemex Director Talks About Oil Reserves," *Foreign Broadcast Information Service*, November 1, 1977, p. M2.

16. "The Future for Mexico," op. cit. p. 6.

17. Susan Kaufman Purcell, *The Mexican Profit-Sharing Decision* (Berkeley: University of California Press, 1975), p. 24.

18. Alan Riding, "Congress in Mexico Will Get New Power," *New York Times*,

---

reserves of 440 trillion cubic feet. Recoverable gas at $2/Mcf net after royalty may reach 150 trillion cu. ft.

(For elaboration, see John A. Masters, "Deep Basin Gas Trap, West Canada," *Oil and Gas Journal*, September 18, 1978, pp. 226–41.)

September 18, 1977, p. 20; "Mexico: Tethered Watchdog," *Latin America Political Report*, October 14, 1977, p. 317.

19. "5,000 Persons Demonstrate Against Oil, Gas Sale to US," *Foreign Broadcast Information Service*, March 21, 1978, p. M1.

20. Alan Riding, "Mexican Concerned that Reliance on Oil May Aggravate Ills," *New York Times*, December 31, 1977, p. 23.

21. Douglas Martin, "Mexicans Count on Oil to Help Repay Debts and Bolster Economy," *Wall Street Journal*, October 26, 1977, p. 1.

22. "President Defends Planned US Gas Pipeline," *Foreign Broadcast Information Service*, October 6, 1977, p. M1.

23. Franz A. von Sauer, *The Alienated Loyal Opposition* (Albuquerque: University of New Mexico Press, 1974), p. 44; Gordon, op. cit., p. 12.

24. *Latin America Political Report*, October 14, 1977, p. 318.

25. John F. H. Purcell and Susan Kaufman Purcell, "Machine Politics and Socioeconomic Change in Mexico," in *Contemporary Mexico*, ed. James W. Wilkie, Michael C. Meyer, and Edna Monzon de Wilkie (Berkeley: University of California Press, 1976), p. 366.

26. "Mexico's Oil, Foreign-Exchange Reserves Soar; President Cites Economic Recovery," p. 6.

27. Loring Allen, *Venezuelan Economic Development: A Political-Economic Analysis* (Greenwich, Conn.: JAI Press, 1977), pp. 304–05.

28. U.S., Department of Commerce, *Venezuela: A Survey of United States Business Opportunity* (Washington, D.C.: U.S. Office of International Marketing, June 1976), p. 43.

29. Howard I. Blustein, *Venezuela, Area Handbook, 1977* (Washington, D.C.: Foreign Advisory Service of American University, 1977).

30. Jorge Díaz Seprano, "Speech before the Chamber of Deputies, October 26, 1977," in *Comercio Exterior de Mexico*, December 23, 1977, p. 483.

7

THE BENEFITS TO
THE UNITED STATES
FROM MEXICAN PETROLEUM

Mexico is located strategically on the southern flank of the United States. This, coupled with the very real domestic as well as bilateral political-economic problems raised by the large and growing influx of illegal Mexican aliens, has caused the United States to develop a legitimate self-interest in promoting the political stability of the Mexican government, more amicable bilateral relations, and rapidly rising living standards, especially for poorer Mexicans.

The fact that Mexico has petroleum reserves sufficient to allow large, steady, year-to-year expansion in the production and export of crude oil and natural gas creates an economic climate conducive to rapidly rising living standards throughout the 1980s. Viewed from the perspective of U.S. self-interest, improved Mexican living standards are especially desirable because they will contribute indirectly to reducing the magnitude of the illegal alien problem and, thereby, help to soothe the principal bone of contention between these neighboring nations. Besides fueling faster economic growth within Mexico, sizable expansions in Mexico's petroleum exports will also enhance Mexico's worldwide status and power. Intercountry relations are never very harmonious when they are between rich and poor, strong and weak. Comparison with the easier relationship of the United States with Canada suggests that an important cause of the latent tension underlying all U.S.-Mexican bilateral relations is the huge asymmetry in wealth and power. Rapidly growing exports of Mexican petroleum should lessen this gap.

Any improvement in Mexico's economic and political prospects will,

The National Security section was co-authored by Robert Fisher; Bruce Mohl co-authored the section on Policies of the United States.

in all probability, redound to the benefit of the United States. Thus, the United States would be justified in making a commitment to encourage Mexico to expand the production and export of petroleum. This chapter advances a much stronger case for promoting Mexico's petroleum industry. Expanded imports of Mexican petroleum will help the United States to achieve three important domestic goals, as well as the aforementioned diplomatic goals: reducing U.S. vulnerability to sudden interruptions in oil supplies, assuring that the United States satisfies its basic energy needs efficiently, and reducing environmental damage.

NATIONAL SECURITY

Prior to the late 1940s, the United States was self-sufficient in crude oil; the Gulf Coast states actually exported large quantities to Western Europe. However, by 1950, rapidly expanding exports from lower-cost Persian Gulf sources had nearly driven U.S. oil out of European markets. Indeed, small but growing amounts of Persian Gulf oil were beginning to be sold in the United States.

Owners and producers of domestic crude oil sought to prevent any significant erosion of their product's market share by persuading the federal government to restrict oil imports. Because it would not be politic to confess selfish motives, they offered a national security justification: if the flow of oil imports was interrupted, the United States would face severe energy shortages until alternative supplies could be developed or until steps could be taken to reduce demand. This would take several years. During the interim, the absence of substitutes would produce economic havoc.

Until the late 1960s, the national security justification for oil import controls rang false. The United States imported relatively small quantities of oil, and prior to 1970, virtually all of it came from what were militarily and politically secure Caribbean (mainly Venezuelan) and Canadian sources. Prolonged interruption from either source was unlikely.

From 1970–73, the security of U.S. oil supplies deteriorated rapidly, as falling domestic production of crude oil and natural gas caused consumption of imported oil to jump at a 30 percent annual rate. At least partly as a result of the sellers' market precipitated by soaring U.S. demand, the leading oil exporting nations within OPEC became more cohesive. Thus, for the first time—short of a war involving the major powers—sizable interruptions in the world oil trade became a real threat. Unfortunately, when members of OAPEC called an embargo in 1973, the United States was ill-prepared.

In the years following the OAPEC embargo, the security of U.S. oil supplies has continued to deteriorate. With the exception of a temporary reversal in 1977–78, attributable to completion of the trans-Alaskan pipeline, U.S. production of crude oil has continued to fall. Moreover, imports from the relatively secure sources of supply in nearby Canada and Venezuela have also fallen substantially. As a result, the United States depends increasingly on more distant sources of oil, such as Iran, Nigeria, and the Arab nations; therefore, sea transport lines have become less defensible.

The apparent lack of control over vital oil supplies has led many to argue that the United States should reduce its dependence on imported oil. Indeed, the first response of the U.S. government to the 1973 embargo was President Richard M. Nixon's proclamation of Project Independence, to enable the United States to meet its energy needs by 1980 "without depending on any foreign sources."[1] The overly ambitious goals of Project Independence were dropped quietly but quickly. Nevertheless, reducing U.S. dependence on imported oil has been the goal of both succeeding presidents.

The basis for the present oil insecurity of the United States is twofold. First, the interests of the small group of countries in control of the bulk of the world's exportable oil production differ from the interests of the United States on a variety of political and economic issues. Second, the United States is becoming more dependent on distant sources of oil at a time when its principal rival, the Soviet Union, is growing in military—especially naval—strength. Increased imports of crude oil and natural gas from Mexico's Reforma and Campeche fields would permit a political, economic, and geographical diversification that would, in part, alleviate these threats and enhance the energy security of the United States.

Diversification from Political Threats

In October 1973, OAPEC's members initiated an oil embargo in an effort to change the Western consuming countries' attitudes toward the Arab-Israeli conflict and to "persuade" them to follow less "pro-Israeli" policies. OAPEC set up a system of categorization that labeled consuming countries as *friendly*, *neutral*, or *hostile* to the Arab cause, and it attempted to allocate oil exports accordingly. The United States and The Netherlands were the primary targets of the OAPEC embargo, as the Arabs hoped they would pressure Israel into returning all of the territories occupied in the 1967 war and to restore sovereignty to the Palestinians. The other Western countries affected by the oil cutback were expected to try to influence the United States to adopt a more pro-Arab stance.

The sudden cutback in OAPEC's oil exports caused political havoc throughout the West. The major, non-Communist powers began to split in their attitudes toward Israel, with the United States maintaining a relatively strong pro-Israeli position. On the sixth day of the October War, the United States began a military airlift to resupply Israel. The airlift lasted 32 days, and thereafter, military supplies continued to be sent by sea. When the Soviet Union suggested it might send military forces into the Middle East, President Nixon ordered all conventional and nuclear forces on military alert.

Because only 17 percent of the U.S. oil supplies were coming from OAPEC sources at the time of the embargo, the United States had the capability to withstand Arab threats. Even though they were *not* the principal targets of the embargo, Europe and Japan—almost entirely dependent upon imported oil—were in a much more vulnerable position. On November 6, 1973, the foreign ministers of the European Economic Community (EEC) agreed on a resolution calling for Israel to withdraw to the lines it had held at the time of the first cease-fire on October 22 and for full implementation of the Security Council's Resolution 242, calling for the end of Israel's occupation of territories conquered in 1967 and for Israeli recognition of the Palestinians' legitimate rights.

Japan, lacking any indigenous energy supplies, was even more vulnerable to oil import restrictions. Japan modified its Middle East policy and also adopted a resolution appealing to Israel to withdraw to the 1967 borders and to negotiate an agreement with the Arabs.

OAPEC's use of the oil weapon succeeded in producing splits and tensions within the Western alliance and the EEC. Most disagreements revolved around the core issue of national security. During the 1973 War, the United States pursued a policy that took account of the U.S. concern for the global balance of power vis-a-vis the Soviet Union and aimed at preventing a Middle East shift in Moscow's favor. The European countries and Japan, concerned primarily about assuring their continued access to oil supplies, were not prepared to endanger their economic vitality. Apart from Portugal, which allowed U.S. planes to refuel in the Azores on their airlift to Israel, European governments refused to lend active support to the pro-Israeli stance of the United States. Great Britain declared an embargo on arms deliveries to all combatants—a policy that affected Israel more than the Arabs. West Germany protested the transfer of U.S. military equipment from Germany to Israel, causing Secretary of Defense James Schlesinger to issue a veiled threat that the United States might reconsider its military presence in Germany.[2]

Within the EEC, nationalist interests prevailed. Because of its supposed pro-Israeli stance, OAPEC had singled out The Netherlands as a target for a total embargo. When the Dutch urged the other EEC countries to exhibit solidarity through oil sharing, Great Britain and France opposed the request. Reportedly under French and British influence, the Oil Committee of the Organization of Economic Cooperation and Development (OECD) decided not to put its oil-sharing system into action.[3] Although this lack of cooperation did little actual economic damage, since available oil was distributed more or less equally by the multinational oil companies, the political damage was considerable. The EEC had to face the harsh truth that national interests had prevailed over European solidarity.

The development and sale of Mexico's petroleum will help to reduce Western dependency on Arab oil. While sharing a common interest with all other oil exporters in wanting to maintain or raise per barrel oil revenues, Mexico is not a member of OPEC and would be unlikely to join OAPEC in either an embargo or a production cutback connected to Middle East politics. The Mexicans have no nationalistic, cultural, political, or religious reasons for supporting an Arab campaign against Israel. Moreover, since Mexico is currently selling oil to Israel, to do so would be unprofitable. While Mexico may at some point refuse to sell oil to the United States or other countries for political reasons (an event the author judges to be unlikely), the motivation for such a refusal is likely to be quite different from OAPEC's.

Insurance companies are profitable because of the fact that the total risks of adverse outcomes can be reduced substantially by diversifying their portfolio of insured activities among several risky, but unrelated, alternatives. In precisely the same way, diversification of oil sources, even if each source is equally insecure, will promote the oil security of the United States. Each barrel of oil the United States buys from Mexico will reduce the residual demand for OAPEC oil by the United States. Thus, a renewed OAPEC embargo would be less harmful and, as a direct corollary, probably less likely. If Mexico chose to halt oil sales to the United States, financial considerations would most likely prompt it to attempt to raise sales to other countries. Because of oil's fungibility, the net effect on the United States would be slight: higher sales of Mexican oil to other countries would simply free more non-Mexican oil for sale to the United States. In short, as long as there are multiple oil sources and only a one-source oil embargo, the net effect would be slightly higher costs due to increased shipping distances.

Technical characteristics of most of Reforma's oil fields require that large quantities of natural gas be produced in association with (that is, as a by-product of) crude oil. But the Mexican economy's ability to consume this natural gas usefully is rather limited—the chief domestic use will be to replace fuel oil as a boiler fuel for electricity generating plants. If Mexico's crude oil production is to continue to expand, large quantities of essentially costless (in the economic sense of near-zero resource costs) natural gas will be available for export. Because of the high cost of liquefaction, the United States offers Mexico its only commercially feasible export market, at least through the mid-1980s. Any surplus natural gas that cannot be exported must be flared. Therefore, if political complications (to be discussed shortly) can be resolved, Mexico has strong financial incentives not only to export natural gas to the United States but to avoid the gas-wasting consequences of an embargo. In sum, the United States should regard the political security of imports of Mexican natural gas as being greater than that associated with imports of Mexican oil and certainly much greater than with imports of OPEC or OAPEC oil.

Perhaps the most important reason expanded exports of Mexico's oil and natural gas would enhance the energy security of all oil importers lies in its psychological value. The OPEC countries are well aware that Mexico has huge petroleum reserves which are growing rapidly as exploration continues. The mere knowledge that there is a growing new petroleum source seems likely to reduce the power of other oil exporters to influence the West with implicit threats of an oil embargo.

Geographic Diversification

Mexico's geographic location enhances the oil security of the United States in two ways. First, Mexico is distant from the numerous sources of conflict that could disrupt the critical flow of oil from the Middle East and Africa. Second, being adjacent to the United States, the vital sea lanes are short and, in the event of conflict, would be relatively easy to defend.

At least six sets of factors can be identified that promote international instability and could lead to conflict situations which would disrupt the critical flow of oil from the Persian Gulf.

First, language, ethnic, and religious differences within the area give rise to tensions and serve as possible sources of conflict. Minority groups, such as the Kurds in Iraq, seek autonomy, while central governments try to maintain and extend control.

Second, differences over continental shelf boundaries may create animosities in the future. New offshore oil discoveries have accelerated the claims of nations bordering on the Persian Gulf (to Arabs, it is the Arabian Gulf), and conflict situations may arise over offshore oil and fish resources, as they have in the Aegean and South China seas.

Third, the stresses of accelerated modernization may cause ruptures in traditional societies. The rapid injection of Western money, thought, and technology into fundamentally feudal societies is a major potential source of future political disruption. The Soviet Union and others may try to exploit this disruption. Parliaments in Bahrain and Kuwait have been dissolved when they called for faster and greater reforms than the rulers were willing to grant.

Fourth, there may be external intervention by either local states or superpowers. The vast oil riches offer a temptation some nations may be unable to resist. Examples of external intervention abound: Iran's aiding Kurdish rebels in Iraq, U.S. arms sales to Iran, Soviet military and economic influence in Iraq, Iranian military intervention against Dhofar rebels in Oman, and the Saudi's growing economic influence in Yemen.

Fifth, Great Britain's withdrawl of military forces from the Persian Gulf has left a power vacuum that has turned into a local arms race. Iran has become the dominant military power within the region. The other states may be reluctant to join a Persian Gulf security pact in which they might be legitimatizing Iranian military intervention. But in the absence of such a pact, a higher level of regional instability can be expected.

Sixth, the Arab-Israeli conflict is a continuous source of possible Persian Gulf hostilities.

Mexico is comfortably distant from these six sources of potential conflict. Disturbances within the Persian Gulf are unlikely to affect Mexico's oil or natural gas production. Mexican oil should continue to flow even if war or civil unrest upsets Persian Gulf exports. Diversification of the sources of oil supplies away from an area as volatile as the Persian Gulf must raise the security of the world's oil flows.

The second benefit attributable to Mexico's geographic location stems from the fortuitous fact that it is adjacent to the United States. Therefore, the risk of interruptions in Mexico's oil exports because of certain types of military actions can be reduced considerably. In order to evaluate these military-security benefits, it is useful to examine briefly five types of military situations.

Nuclear War. A nuclear war is likely to be short but destructive. The outcome of the war would be determined by such factors as each side's stock of nuclear weapons at the outbreak, the targeting strategies of the participants, and the surprise or lack of it in a first strike. While Mexican oil might prove useful to the United States in a postwar recovery, the outcome of such a war is unlikely to be determined by oil and gas considerations. In short, the development of Mexican oil and gas has little relevance to the nuclear war scenario.

Conventional War. Mexican oil could play an important role in U.S. energy security if a major power nonnuclear war continued for more than a few months. The size, distribution, and doctrines of the North Atlantic Treaty Organization (NATO) and Warsaw Pact forces indicate that both sides could expect any European conflict to be short. The outcome is expected to be determined by the size and distribution of prepositioned forces, the warning time before the attack, and whether tactical nuclear weapons are used. Mexican oil would not make a great difference if a European war was short.

Unfortunately, strategists who have prepared for short conventional wars have often been mistaken. If this eventuality was to occur, Mexican oil and gas would be a significant supplement to U.S. energy supplies. Assuming Mexico is not sympathetic toward the aims of the U.S. enemy, the fact that a natural gas pipeline runs through Mexico should pose few additional security problems. Moreover, oil transported from Mexico to U.S. Gulf Coast refineries is much easier to protect than oil that must be shipped from the Persian Gulf or North and West Africa. The distance from Houston to southeastern Mexico is about 800 miles—the entire route can be patrolled by land-based aircraft. In sharp contrast, it is about 12,000 miles to the Persian Gulf, 5,000 miles to Nigeria, and 7,500 miles from Indonesia to the U.S. West Coast. Large parts of the Persian Gulf route are far from U.S. military bases and pass near potential enemies.

Limited War. U.S. strategists must prepare for two types of limited wars: a war directly involving one or more large oil exporters (most likely Persian Gulf countries) and perhaps involving the United States or a war that involves the United States but in which no large oil exporter has a significant interest. The world oil trade is likely to be disrupted if the limited war involves a large oil exporter. Thus, Mexican oil would be especially valuable if this eventuality should occur. The United States continues to produce enough oil domestically to supply the demands of conducting a limited—Vietnam magnitude—war. Thus, if no oil exporters are involved, a limited war poses no direct threat to the energy security of the United States. However, tankers traversing the Persian Gulf–U.S. route could become targets for attack, as they pass near such strategic and potentially volatile places as the African Horn and South Africa. Assuming hostile actions are taken, it would be necessary for the U.S. military either to eliminate its cause or to provide air or naval cover for the tanker fleet. Mexico is close to one potential U.S. enemy—Cuba. However, given the relative naval and air power of the two countries in the Caribbean, it seems unlikely that Cuba would seek to challenge either the United States or Mexico by attempting to interdict the Mexican oil trade.

Undeclared Naval War. The Soviet Union is the only country with a navy capable of contesting U.S. naval supremacy. Recent Soviet adventures in the Horn of Africa, Mozambique, Angola, and West Africa raise the specter that the Soviet Union is trying to acquire naval and air base rights close to the Persian Gulf–North Atlantic shipping route. However, because interference to U.S. shipping would probably result in rapid escalation of the conflict into nonnaval areas and because the Soviet Union's present structure of naval bases makes its distant water fleets vulnerable in any protracted conflict, this does not seem to pose a substantial danger to the Persian Gulf oil trade to the United States.[4]

Terrorism. The coupling of rising terrorism with the proliferation of lightweight, highly accurate weapons systems poses a threat to international oil flows. Wellheads, pipelines, terminals, tankers, refineries, and so forth are all vulnerable to a determined small-scale attack. Terrorists could attempt to disrupt the Mexican–U.S. petroleum trade. However, the shorter routes and the active interest of both countries' governments in preventing this disruption would seem to make Mexican exports relatively more secure.

ECONOMIC EFFICIENCY

World oil prices will be based on Persian Gulf plus freight pricing as long as the Persian Gulf producers continue in their present role as the residual suppliers for all the major oil-importing nations. Because of the long lead times needed to develop large new oil and natural gas supplies, Persian Gulf plus freight pricing seems certain to prevail throughout the 1980s—even if Mexico expands production and exports at the high rates that are technically feasible and economically profitable. Therefore, the United States should not expect to reduce the prices it pays for imported fuels merely because it substitutes Mexican oil and gas for OPEC oil. Nevertheless, rising imports of Mexican petroleum promise three potentially significant economic benefits for the United States: (1) a sizable reduction in total spending for energy, because increased imports of more secure Mexican oil should reduce the need to commence large-scale, commercial production of higher-cost domestic alternatives (such as synthetic gas or oil made from coal, oil shale, nuclear power, and solar energy) in order to ensure an adequate level of energy security; (2) a sharp reduction in capital demands and, as a corollary, less rigid capital stock requirements; and (3) strengthened exports, which will improve the U.S. balance of trade.

Reductions in Energy Expenditures of the United States

The economic dislocations precipitated by the 1973 OAPEC embargo demonstrated that undue reliance on energy supplies that can be interrupted poses a very real threat to the U.S. economy. Recognition of this danger has prompted every president since Dwight D. Eisenhower to adopt policies aimed at reducing U.S. reliance on insecure—nearly always assumed to be imported—petroleum. Two basic approaches are possible: reducing the total petroleum demands of the United States or raising the quantity of secure—nearly always assumed to be domestic—petroleum supplies.

President Jimmy Carter's energy policy has been premised on the assumption—questioned by most independent energy experts—that little can be done to encourage greater domestic production of either crude oil or natural gas. Therefore, it focuses on reducing petroleum demands either by encouraging greater conservation or by encouraging expanded production of those nonpetroleum fuels that the United States has in abundance. Large investments to promote conservation and additional production of alternative, currently commercial domestic fuels—especially coal—are economically profitable at mid-1978 prices. Hence, these investments are being made voluntarily. Unfortunately, there are many large uses (for example, transportation and residential and commercial heating) for which crude oil and natural gas have no presently acceptable substitutes. If the United States is to achieve the target level of oil imports judged by the president to be safe, additional massive investments will be necessary to implement more stringent conservation measures and to produce new types of energy that can be substituted directly for crude oil and natural gas.

Solar power, synthetic oil and gas made from coal, and oil shale are the three petroleum alternatives receiving the most attention from policy makers. Unfortunately, with the exception of small quantities of solar energy that can be exploited through better building design or small-scale, low-temperature water heating, all of these technologies are far more costly than buying imported crude oil at Persian Gulf plus freight prices. Moreover, the technical feasibility of many of the processes necessary for large-scale energy production from these noncommercial sources has yet to be demonstrated. Given the present state of the art, enormous cost overruns and huge amounts of waste appear to be the inevitable by-product of any program to begin pushing large-scale production from these as yet unproven petroleum substitutes. But this creates a real dilemma for U.S. energy policy: either the United States must begin spending immediately many billions of dollars to force large-scale production and consumption of these still uneconomic domestic fuels or it must continue to rely on an unacceptably high level of U.S. oil imports.

The Department of Energy is already spending about $5 billion annually in an attempt to demonstrate the feasibility of large-scale production of alternative fuels, and the president has proposed large tax subsidies to promote solar energy. Nevertheless, to date, there has been little progress toward the commercialization of these petroleum alternatives—substantial production is not anticipated prior to 1990. In short, the United States presently suffers the worst of both worlds:

massive multibillion dollar expenditures are being made in the hope that still exotic forms of energy will prove commercially feasible, and the security of the day-to-day energy supplies of the United States continues to deteriorate.

Because of the relatively greater security of petroleum imports from Mexico, a shift in the composition of the energy imports of the United States in favor of Mexican crude oil and natural gas and away from OPEC oil should offer at least a partial solution to the U.S. oil quandary. Specifically, since Mexico will earn substantially higher profits on every barrel of oil or cubic foot of natural gas sold in the United States (see Chapter 5), it would be in the best interests of both nations to negotiate firm guarantees calling for a large and steadily growing minimum amount of petroleum to be sold by Mexico to U.S. firms each year. With the appropriate guarantees from the United States, Mexico would have absolutely no difficulty raising the capital funds necessary to expand its petroleum output by the agreed upon amount. Conversely, with the appropriate guarantees from Mexico, the United States need not view the supply of Mexican petroleum as being much less secure than supplies of many (strike prone) competitive domestic fuels. As a direct consequence, the United States would be able to reduce the magnitude of its commitment to developing alternative, higher-cost domestic fuels.

Reduced Capital Demands

Assuming current world oil prices and using existing technology, the United States has, basically, three potentially commercial but not yet fully developed sources of energy: coal, crude oil and natural gas from northern Alaska, the outer continental shelf, and from onshore, typically low-productivity reservoirs located in the lower 48 states; and nuclear power. Expensive and time-consuming planning and legal-regulatory hurdles must be cleared before a firm can even begin developing any of these domestic energy sources: land must be gathered, environmental impact statements drafted and approved, and construction permits obtained. In the case of large and controversial proposals, such as developing new coal mines, producing petroleum from the outer continental shelf, building new coal- or nuclear-powered electricity generating plants, and building new pipelines or electricity transmission lines, the regulatory delay tends to be at least three to five years. Then, assuming the project is thought likely to be profitable and all necessary approvals can be obtained, an additional three to eight years and enormous capital investments are necessary before commercial production can begin.

Because southeastern Mexico has prolific reserves of relatively easy-to-produce crude oil and natural gas, the lead times necessary to produce this petroleum would be short, and the capital investments would be small compared with the lead times and capital investments needed to tap U.S. energy sources. Three examples illustrate the dramatic differences:

1. In less than three years and with an investment of about $1.5 billion, Mexico could complete a pipeline able to deliver at least 2 billion cubic feet per day of natural gas to U.S. markets. In contrast, it will take at least six years and cost $10 billion to $15 billion to deliver similar amounts of natural gas from Prudhoe Bay.[5]

2. For an investment of about $2 billion, Mexico expects to be able to expand its output of crude oil by at least 500,000 MBD in 1978. In contrast, the *Oil and Gas Journal* reports that investments in U.S. oil exploration and drilling will total about $10.5 billion in 1978.[6] However, this level of investment will not be sufficient to prevent a small decline in daily output by year-end.

3. It now takes 12-plus years to plan and build a nuclear-powered electricity generating plant, and even approved plants have run into additional unanticipated delays.

High energy-related investment costs place heavy demands on U.S. capital markets; long lead times increase the probability that during the intervening period, changing circumstances will render the investment economically obsolete. Both of these factors result in higher interest rates and, therefore, raise capital costs throughout the economy. These economy-wide cost pressures could be reduced if higher imports of Mexican petroleum were allowed to substitute for higher production of more capital intensive domestic fuels.

The net cost savings for U.S. consumers would be especially dramatic if the United States encouraged large imports of Mexican natural gas. The interstate natural gas pipeline grid of the United States originates in Texas, Oklahoma, and Louisiana and ships natural gas to nearly all states east of the Rockies. But by 1980, domestic natural gas supplies will be sufficient only to allow the interstate pipeline grid to operate just slightly above half capacity (see Map 7.1). The coupling of two facts— that natural gas pipelines enjoy large scale economies and that the Interstate Commerce Commission allows interstate pipelines to charge rates that cover all costs, including a specified maximum rate of profits— entails that customers pay sharply higher transportation costs when natural gas pipelines are half empty. Therefore, substantially lower pipeline charges would be a direct consequence of a decision to allow large amounts of Mexican natural gas to be imported into the United States for sale on interstate markets.

Map 7.1: Capacity of the U.S. Natural Gas Pipeline Grid

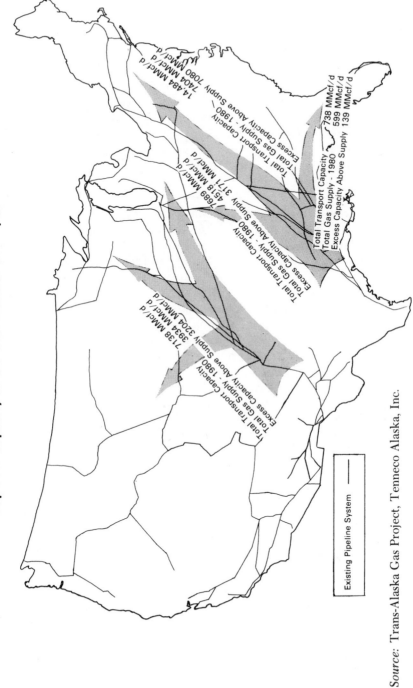

Total Transport Capacity   738 MMcf/d
Total Gas Supply - 1980   599 MMcf/d
Excess Capacity Above Supply   139 MMcf/d

Total Transport Capacity 14,484 MMcf/d
Total Gas Supply - 1980 7404 MMcf/d
Excess Capacity Above Supply 7080 MMcf/d

Total Transport Capacity 7689 MMcf/d
Total Gas Supply - 1980 4518 MMcf/d
Excess Capacity Above Supply 3171 MMcf/d

Total Transport Capacity 7138 MMcf/d
Total Gas Supply - 1980 3934 MMcf/d
Excess Capacity Above Supply 3204 MMcf/d

Existing Pipeline System ——

Source: Trans-Alaska Gas Project, Tenneco Alaska, Inc.

Improving the U.S. Trade Balance

Since mid-1977, Americans have had to pay more for most imported goods because the value of the U.S. dollar has fallen sharply relative to most other major currencies. A large balance of trade deficit is widely believed to be an important cause of the dollar's fall. In this context, an improvement in the U.S. balance of trade would be a valuable joint product of a bilateral agreement guaranteeing that in return for a commitment by Mexico to expand petroleum exports much faster than previously anticipated, the United States would promise to import substantial additional quantities of Mexican natural gas and oil. This contention requires a brief explanation.

A nation's total spending on imports cannot exceed the amount of foreign exchange it has available to spend. Compared with eastern hemispheric oil exporters, such as Kuwait and Saudi Arabia, Mexico's relatively low per capita income and wealth cause it to spend a much larger fraction of its available foreign exchange on importing goods and services; moreover, because of geographical proximity and sharply different levels of industrial and technical sophistication, the United States is far and away the largest exporter of goods and services to Mexico. By agreeing to allow larger petroleum imports from Mexico—especially of natural gas, large quantities of which must be flared if the United States prohibits its importation—the United States could encourage Mexico to raise its petroleum exports to levels higher than they otherwise would be. Adoption of such a strategy would lead to no change in the total bill of the United States for oil imports. However, since compared with any eastern hemispheric OPEC member, Mexico will spend a far higher fraction of any rise in its foreign exchange earnings for U.S. goods and services, a policy that encourages a rise in Mexico's share of the total world oil trade will lead to a net increase in the demand for U.S. exports. Therefore, the net effect of the policy just described would be to reduce the U.S. balance of trade deficit.

THE ENVIRONMENT, PUBLIC HEALTH, AND PUBLIC SAFETY

A major goal of the U.S. energy policy is to reduce the damage to the environment, public health, and public safety that seems to accompany energy production, transportation, and consumption. The following brief profiles of the most important energy types are intended to suggest why an increase in the share of the total energy consumption of the United States provided by crude oil and, especially, natural gas will lead to a reduction in the magnitude of these important problems.

Natural gas is the cleanest-burning fossil fuel—its combustion causes almost no air or water pollution. Moreover, except when it is produced in regions with extremely fragile ecosystems (most notably, the Alaskan North Slope), its production and transportation via pipelines causes modest, generally only temporary, damage to the surrounding environment. Indeed, the only problem with natural gas is that every year, sporadic explosions destroy substantial property and kill several people. However, few observers consider the magnitude of this safety problem to be a cause for substantial public concern.

There are sizable differences in the environmental damage attributable to crude oil of different types (that is, grades) and from different geographic sources. Thus, except for regions like the Alaskan North Slope, the onshore production of crude oil rarely leads to substantial long-term environmental damage. Unfortunately, offshore production of crude oil raises more significant environmental issues. Though oil well blowouts are infrequent, considerable water pollution may occur if one happens offshore. Overseas shipment of oil also raises problems. Modest but chronic water pollution tends to be a by-product of nearly all tanker transportation, and infrequent, but sometimes massive, water pollution can occur when there are tanker accidents. As a result of differences in the chemical composition of crude oils from different regions, there tends to be considerable variance in the air pollution that is a by-product of refining and combustion. With high-sulphur crude oils, these air pollution problems can be severe.

The mining and combustion of coal raises difficult environmental, health, and safety problems. Both deep mining and strip mining are used to produce coal. The ever-present possibilities of mine cave-ins, black lung, and methane poisoning make deep coal mining one of the most dangerous and least glamorous occupations. Huge scars from strip mining and giant piles of waste and overburden have turned large parts of the nation's coal regions into desolate eyesores. Also, water pollution due to acid runoff frequently poisons surrounding lands. There are large differences in the air emissions that are a by-product of the combustion of coal from different regions. Nevertheless, it is generally true that coal combustion emits far more particulates and sulphur oxides than does the combustion of crude oil.

With the possible exception of thermal pollution of water bodies, nuclear power has yet to cause tangible environmental problems. However, only a modest expansion in the size of the U.S. nuclear power industry is likely to be possible due to tremendous controversy over the appropriate answers to the following questions. How safe are nuclear reactors? Can nuclear wastes be stored and transported safely? What are the dangers of nuclear proliferation, terrorism, and blackmail?

As time passes, the U.S. plans to get substantial amounts of energy from oil shale, synthetic oil and gas made from coal, and imports of liquefied natural gas (LNG) from noncontiguous countries. The production of petroleum substitutes from oil shale and coal will be far more disruptive to the surrounding physical environment than the production of crude oil. Expanded LNG use has been widely attacked because of the fear that a catastrophic explosion may occur as a result of either accident or sabotage.

For the reasons just summarized, nearly all authorities rank natural gas (but not LNG) and crude oil as the two preferable forms of energy. Therefore, if because of the greater security of imports of Mexican petroleum, the United States decides that it can increase its total imports of crude oil and natural gas, then there will also be a substantial reduction in the total energy-related damage to the environment and to public health and safety.

POLICIES OF THE UNITED STATES

Policies to promote expanded imports of Mexico's crude oil and natural gas would yield substantial benefits to the United States by encouraging Mexico to expand its petroleum production and by reducing U.S. reliance on oil imports from eastern hemispheric OPEC sources. Judged by economic, environmental, and security criteria, the fact that the United States offers the only potential market for Mexico's natural gas exports suggests two reasons why expanded imports of this product would be especially desirable. First, since Mexico would be unable to divert natural gas shipments to other foreign markets, it would be very costly for it to halt its sales of natural gas to the United States. Therefore, politically inspired interruptions in U.S. supplies of Mexican natural gas are especially unlikely. Second, since Mexico's natural gas is produced in association with (that is, as a by-product of) crude oil, and since Mexico is reluctant to raise its crude oil production as long as the associated natural gas must be flared, one of the most efficacious ways for the United States to encourage Mexico to expand its crude oil production would be by allowing the otherwise worthless associated natural gas to be sold in the U.S. market. As of mid-1978, the U.S. government had done almost nothing to promote the rapid expansion of Mexico's petroleum industry. Even worse, the United States was pursuing energy policies that actively discouraged large imports of Mexican natural gas.

The Gasoducto

In early 1977, after a series of discoveries on both the eastern and western flanks of the main Reforma Trend, it was evident that southeastern Mexico's reserves of associated natural gas were far larger than previously anticipated. The *Oil and Gas Journal* described the causes and dimensions of the changed picture as follows:

> Those structures (Agave to the east and Mundo Nuevo, Paredon, Cacho Lopez, Copano, Giraldas, and Sunuapa to the west) contain either volatile oil or gas and condensate. Their gas-oil ratios range from 3,500 [cubic feet] :1 [barrel] to 7,000:1, compared with the average 1,000:1 for fields in the central portion of the play. . . .
>
> A conservative estimate puts the associated gas output at nearly 5 billion cfd [cubic feet per day] by 1980, when Chiapas-Tabasco will be producing about 2,000,000 b/d of oil. In its original 6-year plan, unveiled by Pemex late in 1976, the government-owned agency had forecast gas production at only 4 billion cfd for the entire country by 1982.[7]

If Reforma's crude oil is to be produced, the associated natural gas must be produced. In 1977, in order to prevent wasteful flaring of this associated gas, Pemex proposed building a giant 840-mile pipeline (or "gasoducto") that would link the oil fields of Chiapas-Tabasco with the industrial center of Monterrey and would, by early 1980, connect with the U.S. natural gas pipeline grid at the Texas border near Reynosa (see Map 7.2).[8] The 48-inch diameter gasoducto was to be the largest natural gas pipeline ever laid in the western hemisphere—its designed capacity was 2.7 billion cubic feet per day, equivalent to nearly 400,000 barrels per day of crude oil, and the estimated cost of delivering Reforma's gas to the Texas border was only \$.40 per Mcf.[9] Mexican industries were expected to be able to consume only about one-quarter of this natural gas. However, Pemex anticipated that any surplus—estimated at 2 billion cubic feet per day—would be readily marketable within the gas-hungry United States. Assuming daily exports of 2 billion cubic feet at the \$2.60 per Mcf price sought by Pemex, the gasoducto would have been fully paid for after just one year's operation. Even at the \$1.75 per Mcf. price proposed for new domestic natural gas in President Carter's National Energy Plan, the payback period would have been less than 20 months. Because of the excellent economics and the enormous potential for earning new foreign exchange, the IMF exempted any funds borrowed for the gasoducto from its restrictions on Mexico's net foreign borrowings. Private lenders responded eagerly.

In August 1977, Pemex signed letters of intent to sell natural gas to six U.S. interstate natural gas pipelines.* Under the terms of the letters of

---

*The companies were Tenneco, 37.5 percent share; Texas Eastern Transmission, 27.5 percent share; El Paso Natural Gas, 15 percent share; Transcontinental Gas Pipeline, 10 percent share; Southern Natural Gas, 10 percent share; and Florida Gas Transmission, 10 percent share.

Map 7.2: Route of Pemex's Gasoducto

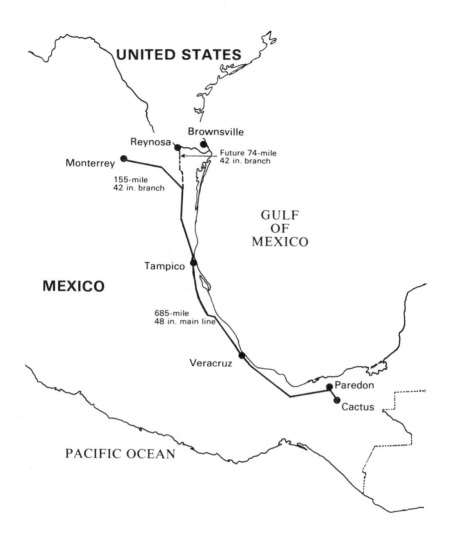

UNITED STATES

Brownsville

Reynosa

Future 74-mile
42 in. branch

Monterrey

155-mile
42 in. branch

GULF
OF
MEXICO

Tampico

MEXICO

685-mile
48 in. main line

Veracruz

Paredon

Cactus

PACIFIC OCEAN

Source: Oil and Gas Journal, June 5, 1978.

intent, Pemex would begin immediately to deliver 50 million cubic feet of natural gas per day from fields located directly across the Texas border. But upon the gasoducto's 1980 completion, Mexico's daily natural gas shipments would soar to 2 billion cubic feet. The U.S. companies agreed to pay $2.60 per Mcf. and there would be further price hikes if heating oil prices were to rise.

The proposed Mexican gas deal was unacceptable to President Carter. His National Energy Plan called for fixing the price of new domestic natural gas at $1.75 per Mcf. and U.S. gas transmission companies were paying only $2.16 per Mcf. for natural gas imported from Canada. The following statement by George P. Mitchell, chairman of the 3,500-member Texas Independent Producers and Royalty Owners Association, illustrates why acceptance of the $2.60 price for Mexican natural gas would have made adherence to these other prices politically untenable. While supporting the idea of Mexican gas imports, Mitchell said:

> What we're talking about here is subsidizing a foreign producer at the expense of domestic producers. We now have a system in which producers cannot receive adequate incentives to develop the nation's gas resources to the fullest, but one that encourages wasteful usage of gas. So now we're in a position of desperately needing imports, for which we'll pay more than if we had developed our own resources without price controls.[10]

In addition to an unacceptably high price, the Mexican gas deal had a second major defect. Because the pricing formula linked future prices for Mexican gas directly to the price of imported fuel oil, the effect of the agreement was to indirectly peg prices for Mexican gas to the price of imported crude oil. Acceptance of this principal would have undercut the administration's rationale for maintaining price controls of domestic natural gas and crude oil. Faced with these problems, the Carter administration believed its only recourse was to kill the proposed Mexican gas deal. Thus, the Department of Energy refused to authorize the six gas transmission companies to purchase Mexican natural gas.

Because of Mexico's long history of rancorous relations with the United States over energy matters, Mexico's President Portillo had been criticized severely for supporting the gasoducto, a scheme that critics compared unfavorably to the Panama Canal. Portillo defended his position by pledging that Mexico would receive a "fair" price for whatever gas it would be unable to store or consume domestically.[11] Mexico could earn profits if it was to export its natural gas to the United States at any price greater than $.40 per Mcf. Nevertheless, in view of President Portillo's public commitments and the two facts that six U.S. companies had voluntarily agreed to pay at least $2.60 per Mcf. and that in early

1978, the U.S. Department of Energy appeared to approve, in principle, a deal to pay $3.42 per Mcf. for LNG from Indonesia, it was politically untenable for Mexico to accept less than $2.60 on its U.S. sales. In short, the U.S. decision not to allow imports of Mexican natural gas at the $2.60 price has led to a sharp deterioration in U.S.-Mexican energy relations. Relations are likely to worsen as long as Mexico is forced to flare ever-larger amounts of natural gas because the United States has closed its borders to Mexican imports.

There are no villains in the gasoducto story: domestic constraints forced both presidents to adopt positions from which compromise will prove difficult. However, the issue certainly illustrates one of the flaws in the continued commitment of the United States to natural gas price controls.[12] Crude oil and natural gas are close substitutes. But because it is much cheaper to transport, crude oil is the more valuable product, that is, at the wellhead, crude oil will sell for a higher price than an equivalent amount of natural gas. In early 1978, imported crude oil cost about $14 per barrel delivered to Texas refineries. Since 7 Mcf. of natural gas has an energy content equal to about one barrel of crude oil, Mexico would have been unable to sell its natural gas for more than $2.00 per Mcf. if the United States had not been enforcing domestic price controls. More precisely, the U.S. gas transmission companies were willing to pay $2.60 per Mcf. for Mexico's natural gas only because of the conjunction of two U.S. policies. First, price controls on interstate natural gas were holding the average price to less than $1 per Mcf.; low interstate prices caused the shortages of domestic natural gas. Second, other regulations required gas transmission companies to practice "rolling-in," that is, higher-cost imported natural and synthetic gas are combined with lower-cost domestic gas; the customer's price is based on the combined product's average costs. As long as the rolled-in price is less than the price for an equivalent quantity of crude oil, the gas transmission companies find it profitable to import natural gas, even if the price is higher than the price of crude oil.

If the United States had not been enforcing natural gas price controls during 1977, no gas transmission company would have been willing to pay more for Mexican natural gas than the price of an equivalent amount of imported crude oil delivered to the Texas market. Thus, regardless of the assessment by the two presidents (or anyone else) as to what is the "fair" price, the maximum price for Mexican natural gas sold in the United States would have been somewhat less than $2 per Mcf. in late 1977. If market forces rather than government policies had been allowed to determine this price, it seems likely that a natural gas deal beneficial to both countries would have been consummated in late 1977 and that presidential disagreements over what constitutes a 'fair' price would not have led to deteriorating bilateral energy relations.

Coal Conversion Legislation

A basic goal of President Carter's National Energy Plan is to promote the substitution of coal for oil and natural gas as fuels for electricity generation and large-scale industrial boilers. According to the National Energy Plan:

> Energy resources in plentiful supply must be used more widely, and the nation must begin the process of moderating its use of those in short supply. Although coal comprises 90 percent of domestic fossil fuel reserves, the United States meets only 18 percent of its energy needs from coal. Seventy-five percent of energy needs are met by oil and natural gas although they account for less than 8 percent of U.S. reserves. This imbalance between reserves and consumption should be corrected by shifting from oil and gas to coal and other domestic energy sources.[13]

Three types of criticism have been raised by opponents of the president's proposed legislation. First, expanded production and consumption of coal will exacerbate the environmental and safety problems discussed previously. Second, there is some doubt about the ability of the coal industry to meet the production targets set by the president: productivity in deep mines has dropped from 15.6 tons per day in 1969 to only 8.5 tons in 1977.* Over the same period, the inability to resolve land use issues has set back plans to expand coal production from western strip mines. A long history of labor unrest—best typified by the prolonged winter 1977–78 coal strike—illustrates that there are no guarantees that the supply flow, even from domestic sources, is invulnerable to costly interruptions. Third, because prevailing price differences already create strong economic incentives to switch to coal whenever feasible, the proposed legislation is at best redundant and at worst, if it forces costly retrofitting of plants presently fueled by either oil or natural gas, enormously wasteful. The potential for large U.S. imports of Mexican natural gas gives added weight to each of these criticisms.

The policy of the United States to promote coal conversion is not directed against Mexico. Nevertheless, to the extent this policy is successful, it must have a deleterious effect on the development of Mexico's petroleum industry, since it will reduce the size of the potential market

---

*The productivity drop is due partly to more stringent safety practices and partly to greater numbers of inexperienced miners. However, the General Accounting Office states that deteriorating labor-management relations might be the most important consideration. (See *Wall Street Journal*, March 16, 1978, p. 20.)

for Mexico's natural gas exports. Adoption of a coal conversion policy would raise the unpleasant possibility that even if the United States and Mexico could resolve their pricing disagreements, Mexico would be unable to sell its natural gas to the large industrial and electric utility users of the United States. Should this happen, there would be unnecessary damage to the U.S. environment and public health and, since the nation's interstate natural gas pipeline will be operating at only about half its capacity, unnecessary economic waste.

Import Policy

All recent U.S. presidents have been deeply concerned about the high level of oil imports needed by the United States if it is to satisfy its total oil demands. To help remedy this problem, President Carter urged Congress to pass a high tax on domestic crude oil, thereby raising fuel prices and encouraging greater conservation. As of mid-1978, Congress showed little taste for adopting the president's policy. Reluctantly, the president has countered congressional inaction by periodically threatening to take administrative action to impose either tariffs or quotas on imported oil.

The principal reason for the president's reluctance to impose oil import controls appears to be that in order to avoid disproportionately high fuel prices in those regions of the country which consume proportionately more imported oil, especially in the northeast, additional complicated allocation regulations would be necessary. However, the proposed import restrictions suffer from a second defect: the highest costs would be borne by the most secure sources, those western hemispheric countries—Canada, Mexico, and Venezuela—that export proportionately more of their oil to the United States. As with so many other domestic energy problems, the first defect of oil import restrictions could be remedied simply by eliminating crude oil price controls.* The second defect could actually be turned into a virtue simply by grafting to any program of oil import restrictions a system of preferences for oil imports from especially secure (for example, contiguous, western hemispheric, or non-OPEC) sources. Precisely because it would promote the vital security interests of the United States, such a proposal was actually recommended in 1970 by a majority of President Nixon's cabinet task force on Oil Import Controls.[14]

Speedier development of the Mexican oil industry would almost

---

*If crude oil price controls were eliminated, the price of domestic oil at any location would be equal to the delivered price of foreign crude. Thus, U.S. consumers would be indifferent with regard to the two products.

certainly benefit the United States. Yet, as of mid-1978, the U.S. government had done nothing to encourage this result. Even worse, though not specifically designed to do so, some U.S. policies actually have set back the development of the Mexican óil industry. A policy to grant modest (perhaps $.25 per barrel) tariff preferences on oil imports from contiguous (or western hemispheric), non-OPEC countries would offer tangible evidence of the friendship of the United States toward Mexico and of its commitment to encouraging speedy development of the Mexican oil industry.

The Indifference of the United States

Because of the power asymmetry and the bitter legacy from past U.S. exploitations, U.S.-Mexican relations have never been easy. Nevertheless, all observers believe that the prospects for improved relations have risen dramatically since Portillo's accession to the Mexican presidency. President Portillo provides a sharp contrast to his outspoken predecessor, Luis Echeverría. Echeverría was known in the United States primarily for his fiscal irresponsibility and his anti-American attacks. As the first finance minister to be elected president, Portillo appreciates the economic benefits to be gained by developing his country's oil and gas reserves quickly. In endorsing Pemex's expansion plans, President Portillo warned: "This opportunity will come only once in history. We have to transfer a nonrenewable resource into a permanent source of wealth to meet our current needs and those that will come with the increase in our population."[15] President Portillo also appears willing to permit U.S. help to achieve this end: he has invited representatives of many major energy companies to meet with Pemex, and he has made several goodwill gestures toward the United States, including the supply of emergency oil and natural gas during the winter 1976–77 natural gas shortage.

President Portillo's moderate record and constructive actions demonstrate that he is a valuable friend of the United States. His efforts to promote a pragmatic Mexican energy policy merits U.S. attention and support. This will require the energy policy makers of the United States to hew to a fine line in their dealings with Mexico. On the one hand, because fears of U.S. exploitation run deep among most Mexicans, U.S. policy makers should be especially careful not to place Portillo in a position in which he would appear to be succumbing to U.S. pressure. The appearance of yielding to U.S. pressure would cause erosion of Portillo's domestic power base and would be likely to jeopardize future U.S. participation in the development of Mexican oil and gas. On the other hand, recent events suggest that a U.S. energy policy that takes no account of Mexico's petroleum potential will delay the development of

Mexico's petroleum industry and will prove unnecessarily costly for the United States.

Shortly after assuming office, President Portillo paid an official visit to the United States. In a speech to the House of Representatives calling for U.S. help and understanding during Mexico's development efforts, he was interrupted by applause only once—when he began a sentence by saying, "Mexico must solve its own problems."[16] After the applause subsided, President Portillo continued, "Mexico must solve its own problems, and you must examine those of your decisions which may adversely affect or undermine our development effort."[17] The incident reveals much about U.S.-Mexican relations under Portillo. Mexico would like to solve its problems independently, but it realistically accepts the fact that whether intentional or not, the policies of the United States will have an enormous impact on Mexico. Elsewhere, he has elaborated, almost poetically, on this theme:

> The United States must realize the real meaning of the strength of the strong. A phrase, a comment can do a great deal of harm to a weak country and on the other hand, a gesture of understanding and sympathy can do a weak country a great deal of good. I know that it is very difficult to be strong. It is also very difficult to be weak.[18]

When compared with the realism expressed by Mexico's president, the actions of the U.S. government make it look something like a giant with its head in the clouds—persistently and stubbornly ignoring the enormous potential of Mexican petroleum. Even more unfortunate, press reports suggest that some politically insensitive policy makers believe that it is in the interest of the United States to adopt a slightly bullying approach. Thus, the *Los Angeles Times* reported on May 8, 1978:

> After two days of talks in Mexico, high-ranking State Department officials said they do not believe that Mexico can make good on its threat to find other customers for natural gas if the United States does not meet its asking price. A senior U.S. official in the traveling party of Secretary of State Cyrus R. Vance said the Administration thinks that Mexico has little choice but to sell its gas to the United States. The Administration will be "tough but fair" when negotiations begin, the official said.[19]

Similar comments have been attributed to Department of Energy Secretary Schlesinger.[20] Perhaps the most lucid criticism of the U.S. position toward Mexican petroleum appeared in a *Wall Street Journal* editorial entitled "Reviewing the Energy Debacle." The *Journal* writes:

> The second huge cost has been the utter derangement of our relationship with Meixco. Because it embarrassed the lobbying effort with a

$2.60 price, a deal to import huge quantities of Mexican gas was busted by the administration. The Mexicans understandably resent this, as they must resent the current claims that, oh well, they'll come back and sell it to us anyway. . . .

Beyond that, the resentment adds to a growing foreign-relations problem with Mexico. Here we have a nation of 65 million, growing rapidly. It has justifiable grievances against the U.S. on subjects like tomato tariffs and Colorado River water. Illegal immigration is an irritant, building up a huge Mexican minority in the U.S., much of it in areas where Mexico has irredentist interests. With the wrong change of government, this mixture could evolve into a pressing problem of national security.

The Carter energy proposals were intended to solve the problem of energy supplies, and to protect the national security. It will be ironic, though it seems entirely likely, if their most lasting consequence turns out to have been missing the key chance to cement relations with Mexico, head off a potential security problem and insure easy access to our most readily available supplies of energy.[21]

In short, the United States must adopt a more pointedly constructive policy toward Mexico if both nations are to reap the benefits made possible by the existence of huge quantities of Mexican petroleum.

NOTES

1. *New York Times*, November 8, 1973, p. 1.

2. Hans Maull, *Oil and Influence: The Oil Weapon Examined* (Adelphi Paper, no. 117) (London: International Institute for Strategic Studies, 1975), p. 9.

3. Ibid.

4. Geoffrey Kemp, "Scarcity and Strategy," *Foreign Affairs*, January 1978, p. 401.

5. Howard M. Wilson, "Plans Shaping Up for Alaska Gas Line," *Oil and Gas Journal*, April 3, 1978, p. 43.

6. "Industry Spending in U.S. to Hit Record $28.9 Billion," *Oil and Gas Journal*, February 20, 1978, p. 64.

7. "Mexico Speeds Work on Huge Gas Line," *Oil and Gas Journal*, June 5, 1978, p. 121.

8. For a detailed examination of the gasoducto controversy, see Richard R. Fagen and Henry R. Nau, "Mexican Gas: The Northern Connection" (Paper delivered at Conference on the United States, U.S. Foreign Policy and Latin American and Caribbean Regimes, Washington, D.C., March 27–31, 1978).

9. Ibid., p. 122.

10. "Tipro Wants to Intervene in Mexican Gas Case," *Oil and Gas Journal*, September 12, 1977, p. 58.

11. Douglas Martin, "Mexicans Count on Oil to Help Repay Debts and Bolster Economy," *Wall Street Journal*, October 26, 1977, p. 1.

12. For a discussion of other problems with the natural gas price controls, see Richard B. Mancke, *The Failure of U.S. Energy Policy* (New York: Columbia University Press, 1974), chap. 7, pp. 106–21.

13. U.S., Executive Office of the President, *The National Energy Plan* (Washington, D.C.: Government Printing Office, 1977), p. 30.

14. U.S., President, Cabinet Task Force on Oil Import Controls, *The Oil Import Question* (Washington, D.C.: Government Printing Office, 1970), p. 137.

15. "Mexico Grapples with Its Oil Bonanza," *New York Times*, May 7, 1978, p. F-3.

16. "Mexican Chief, in Speech to House, Asks More 'Sensible' U.S. Policies," *New York Times*, February 18, 1978, p. A-12.

17. "Address to the U.S. Congress by José López Portillo, President of Mexico," advertisement in *New York Times*, February 18, 1977, p. A-14.

18. "Mexican Leader, Outlining Plans, Sees a Challenge in National Crisis," *New York Times*, February 1, 1978, p. 4.

19. *Los Angeles Times*, May 8, 1978.

20. "Mexico to Wait Rather Than Cut Gas Price to U.S.," *New York Times*, January 6, 1978, p. D-5.

21. "Reviewing the Energy Debacle," *Wall Street Journal*, October 19, 1978, p. 22.

8

MACROIMPLICATIONS
OF MEXICO'S PETROLEUM

Since the 1973–74 OAPEC oil embargo, the world's energy problems have been widely proclaimed to be the most important and difficult that we will encounter in our lifetime. Two interrelated themes underlie much of the popular and official concern. First, because of voracious energy consumption—especially by the profligate developed countries—the world is rapidly exhausting its finite reserves of crude oil and natural gas. Adherents to this position maintain that unless the principal oil-consuming countries adopt stringent conservation measures immediately, the world's petroleum resource base will be inadequate to support consumption of the magnitude projected by the late 1980s and early 1990s. Second, because Saudi Arabia owns most of the world's known surplus capacity of crude oil, its power over the world oil trade will continue to grow, thereby causing a corresponding growth in Saudi influence and power.

This chapter discusses how a single phenomenon—the on-going discoveries, just since 1972, of 100-plus billion barrels of commercially producible petroleum reserves in southeastern Mexico—discredits the first theme. Moreover, if Mexico chooses to expand its petroleum production rapidly, Saudi Arabia is likely to experience a substantial decline in its power during the 1980s.

POSTPONEMENT OF DOOMSDAY

Malthusian prophecies that the world will suffer tremendous economic dislocations because of the physical exhaustion of an important, but unfortunately finite, natural resource have a long, distinguished (in terms

---

The research for this chapter was conducted by Arthur Boley.

of public support), but not very accurate history. Edward Mitchell has written:

> If the age of an idea contributes to its validity than the doomsday thesis has a lot going for it. However, the doomsayers have not only been consistently vocal, they have also been consistently wrong. America has had less than a dozen years' supply of oil left for a hundred years. In 1866 the United States Revenued Commission was concerned about having synthetics available when crude oil production ended; in 1891 the U.S. Geological Survey assured us there was little chance of oil in Texas; and in 1914 the Bureau of Mines estimated total future U.S. production at 6 billion barrels—we have produced that much oil every twenty months for years. Perhaps the most curious thing about these forecasts is a tendency for remaining resources to grow as we deplete existing resources. Thus, a geologist for the world's largest oil company estimated potential U.S. reserves at 110 to 165 billion barrels in 1948. In 1959, after we had consumed almost 30 billion of those barrels, he estimated 391 billion were left.[1]

The poor predictive record of the Malthusian analyses results from two principal defects. First, suppose the earth is rapidly exhausting its finite reserves of selected natural resources. Then, their relative prices will rise, encouraging larger investments both to increase supplies—by searching for and developing higher-cost sources of the resource and by switching to less expensive, but presumably less efficient, substitutes—and to decrease demands by implementing measures to improve consumption efficiency. Of course, the higher the price rise, the stronger both of these effects will be.

Sophisticated Malthusians admit the logic of the above criticism. But they rebut its relevance by arguing that the necessary price adjustments will be so large and rapid that enormous economic disruption will result. The rebuttal sounds persuasive, but it is purely hypothetical.* After adjusting for general price inflation, there is no recorded instance wherein the price of a natural resource that is vital to the functioning of the world economy has risen rapidly and permanently to a new, much higher level because of the physical depletion of the world's reserves. Though repeatedly prophesied, this lack of evidence of imminent physical shortages suggests the second defect with the Malthusian analyses: they systematically understate the magnitude of the world's finite base of resources producible at or below current prices.

---

*Events following the 1973–74 OAPEC embargo do demonstrate that soaring prices of a vital natural resource can reap economic havoc. However, these price hikes were the result of a political-economic decision by a few major sellers to cut back sales. Since they had (and still have) substantial excess capacity, it was not due to exhaustion of their physical reserves.

The cause of the Malthusians' understatement of the amount of natural resource reserves producible at or below present prices is an ahistorical view of the way in which new supplies of natural resources are discovered. The Malthusians tend to view the world's natural resource base as the largely static quantity thought likely, at any specified time, to be commercially producible from known producing regions. In the specific case of petroleum, the discoveries of large new reserves in virgin areas, such as the Alaskan North Slope, the North Sea, and southeastern Mexico, are the most recent instances of the inappropriateness of the Malthusians' static world view. The petroleum reserves presently thought to be recoverable from just these three new producing regions—the first of which was discovered in 1967—exceeds OPEC's total crude oil production over the entire 18-year period from its founding in 1960 through 1978. If, as now seems likely, additional exploration results in the continued rapid expansion of the geographical dimensions of southeastern Mexico's prolific oil play, net additions to Mexico's known, recoverable petroleum reserves alone will continue to more than offset OPEC's total production for the next several years.

The large petroleum discoveries in northern Alaska, the North Sea, and southeastern Mexico share two important characteristics. First, prior to the successful completion of their first discovery wells, most informed observers did not believe that these regions would be the source of prolific petroleum reserves. Second, in all three instances, large oil reserves had been discovered and development was underway prior to the sharp jump in world oil prices that occurred in 1973–74. Therefore, the discovery and development of these enormous new supplies should not be credited to the post-OAPEC embargo rise in world oil prices.

To date, the petroleum potential of vast areas of the world remains to be explored. As the geologist Bernardo Grossling delights in pointing out, only 23 percent of the world's drilling has been done outside the United States, and 95 percent of the continental shelf is totally unexplored.[2] Higher crude oil prices since 1973–74 increase the attractiveness of searching for oil in these still-virgin areas. The unexpected huge discoveries in Mexico, the North Sea, and Alaska support the inference that future exploration will yield further large additions to the world's stock of petroleum reserves producible at or below present (in real terms) world prices.

MEXICAN PETROLEUM AND THE LIMITS OF OPEC AND SAUDI POWER

The power enjoyed by the 13 member nations of OPEC derives from the fact that worldwide petroleum demands are far in excess of the supplies available from non-OPEC sources. In short, the OPEC countries are the residual suppliers of the world's crude oil. Hence, by acting together to restrict their total exports, these 13 nations have the power to set world crude oil prices at whatever level they desire. Because it has the largest presently known reserves, the most excess capacity, and relatively modest domestic revenue needs (compared with its total oil sales), Saudi Arabia exerts the most economic power within OPEC.

Table 8.1 is derived directly from OECD 1977 projections of the world oil trade in 1980 and 1985. The OECD's projections are similar to those from other respected sources. These projections show that depending on the extent to which the major oil-consuming countries push policies to promote either faster production of indigenous fuels or greater energy conservation, net demand for imports of OPEC oil will range between 24.5 and 35.1 MBD in 1985. This can be compared with actual OPEC exports of nearly 30 MBD in the fourth quarter of 1977.[3] Because most of

TABLE 8.1

OECD's 1977 Projections of World Petroleum Trade
(millions of barrels per day)

Area or Country	1980	1985	
		Reference Case	Maximum Conservation and Indigenous Production
Canada	0.8	1.1	0.7
United States	9.3	9.7	4.3
OECD Europe	12.4	14.7	11.0
Japan	6.9	8.7	7.6
Australia/New Zealand	0.6	0.8	0.7
Centrally planned Europe	−0.5	0.4	0.4
Centrally planned Asia	−0.5	−1.2	−1.2
Oil-importing developing countries	2.9	3.0	3.0
Other countries	1.1	1.2	1.2
Non-OPEC oil-exporting developing countries	−3.0	−3.8	−3.8
Statistical difference	0.5	0.5	0.5
Total net demand for OPEC's oil exports	30.6	35.1	24.5

Note: Net imports (+); net exports (−).

Source: Organisation for Economic Cooperation and Development, *World Energy Outlook* (Paris: OECD, 1977), Table 1, p. 9.

the additional conservation and supply measures endorsed by the OECD are unlikely to be introduced in the near future, the consensus projection of most authorities is that OPEC's 1985 exports are most likely to be near the upper bound of this range.

Closer examination of Table 8.1 reveals one potentially serious defect with the OECD's "reference case" consensus projection that OPEC's oil exports will total about 35 MBD in 1985. Table 8.1 assumes that between 1980 and 1985, the net petroleum exports of all of the non-OPEC oil-exporting developing countries (these include Bahrain, Oman, Brunei, Trinidad and Tobago, Angola, Congo, Egypt, Tunisia, Zaire, Bolivia, Malaysia, Syria, Peru, Thailand, and Mexico) will only rise by 800,000 barrels per day. But net petroleum exports from these non-OPEC sources (even excluding Mexico) are already expanding. If, during the early 1980s, Mexico simply chooses to continue to develop its oil and gas reserves at 1978 rates, its petroleum exports will rise by about 2 MBD between 1980 and 1985. Moreover, if the major oil-importing countries, particularly the United States, can find ways to assist Pemex's efforts to expand petroleum production (for example, by providing technical assistance and markets for natural gas), then Mexico's petroleum exports could rise by as much as 5 MBD between 1980 and 1985. In sum, because the absolute growth in Mexico's petroleum exports between 1980 and 1985 is very uncertain, all projections of OPEC's 1985 oil exports are uncertain. However, because present projections implicitly assume almost no growth in Mexico's petroleum exports during the early 1980s, there is reason to infer that if the projections of OPEC's exporting 35 MBD in 1985 are in error, it will be because they are too high.

Suppose Mexico and the United States both choose to adopt policies to encourage rapid development of Mexico's petroleum industry. Then, OPEC's 1985 oil exports are likely to be closer to 30 MBD than to 35 MBD. What would this mean for OPEC and for Saudi Arabia?

The OECD estimates that OPEC's productive capacity was 38.5 MBD in 1976 and that the consumation of "present expansion plans in OPEC countries will probably increase capacity up to about 45 MB/d by 1980."[4] Being conservative, that is, by assuming no further capacity expansion is undertaken and that OPEC's domestic consumption rises from 2.3 MBD in 1980 to 3.5 MBD in 1985, the OECD projects that OPEC's shut-in capacity will be about 6 MBD in the 1985 reference case. Thus, other things being equal, if Mexico was to raise its petroleum exports by 5 MBD, OPEC's shut-in capacity in 1985 would nearly double, to about 11 MBD. If they are to maintain real prices at their 1977 level, the cartel members would have to allocate substantial production cutbacks. Arguments over the "fair" distribution of these cutbacks seem likely to lead to some diminution of the cartel's power.

Most assessments of OPEC's future behavior begin by dividing its members into at least two groups: low absorbers and high absorbers. Relative to their present oil sales, the low absorbers—Saudi Arabia, Kuwait, Qatar, the United Arab Emirates, and Libya—have small populations and small revenue needs. Their current accounts are nearly always in surplus, causing them to accumulate large reserves of foreign exchange. In comparison, the high absorbers have much larger populations and much lower per capita oil revenues. Because they are in the midst of expensive programs to modernize and industrialize their economies, most high absorbers either already are or soon will be running current account deficits. Hence, they are thought likely to be rather more reluctant about reducing their oil exports. If this orthodox analysis is correct, then the low absorbers—especially Saudi Arabia—will have to absorb most of the cutback in exports due to any rise in Mexico's foreign petroleum sales.

Table 8.2 presents the OECD's estimates of OPEC's total productive capacity and actual production in April 1976 and its projections of OPEC's total productive capacity in both 1980 and 1985. If the high absorbers wish to produce near capacity and if because of the rapid expansion of Mexico's petroleum exports, the export market for OPEC's oil is only 30 MBD, the five low absorbers would be able to export a total of only about 7.5 MBD in 1985. Table 8.3 shows the oil production of the OPEC countries during the fourth quarter of 1977, when OPEC's total domestic consumption was less than 2 MBD. At that time, the five low

TABLE 8.2

OECD's Estimates of OPEC's Production Capacity and Utilization (MBD)

Country	April 1976 Capacity	April 1976 Production	1980–85 Capacity
Saudi Arabia	11.5	8.3	15.0
Kuwait	3.6	1.8	3.5
Iraq	3.0	1.5	6.0
Iran	6.5	5.5	7.0
Libya	2.5	1.9	2.5
Others	11.0	9.5	11.0
Total	38.5	29.0	45.0

Source: All data are derived from Organisation of Economic Cooperation and Development, *World Energy Outlook* (Paris: OECD, 1977), Table 28, p. 87.

absorbers were actually producing more oil (15.87 MBD) than the eight high absorbers (15.64 MBD), and Saudi Arabia alone was producing 8.86 MBD. This suggests it is unlikely that the low absorbers would voluntarily accept a sharp cutback in oil exports. Thus, it seems likely that if Mexico becomes a large petroleum exporter during the 1980s, most of the OPEC countries will have substantial excess capacity. This, in turn, should lead to some pressure on world oil prices and a reduction in the cartel's economic power.

Based on the preceding analysis, some might infer that encouraging Mexico to expand its petroleum production offers a tactic for destroying OPEC. Such an inference would be naive and almost certainly mistaken. Mexico is neither a member of OPEC nor would it seem to be in its self-interest to accept membership were it offered. More precisely, by remaining outside of OPEC, Mexico can sell all of the oil it desires at the Persian Gulf price plus freight. Unlike OPEC's members, who are burdened with the responsibility of setting and maintaining world oil

TABLE 8.3

Production by OPEC's Members During 1977 Fourth Quarter (MBD)

Country	Production
Saudi Arabia	8.860
Kuwait	2.078
Neutral zone (production shared by Saudi Arabia and Kuwait)	0.399
Qatar	0.491
United Arab Emirates	1.991
Libya	2.047
Subtotal low absorbers	15.866
Iran	6.000
Iraq	2.400
Algeria	1.100
Nigeria	1.914
Gabon	0.225
Indonesia	1.685
Venezuela	2.125
Ecuador	0.183
Subtotal high absorbers	15.632
Total	31.498

Source: The Oil and Gas Journal, May 29, 1978, p. 21.

prices, Mexico would have no need to hold production substantially below capacity. Nevertheless, it certainly would not be in Mexico's self-interest to produce so much oil as to topple the OPEC cartel. If this were to happen, Mexico's oil revenues would plunge sharply, as world oil prices plummeted to competitive levels.

Mexico's energy policy makers are fully aware both of their relative advantages vis-a-vis OPEC and of the high losses Mexico would suffer in the event of an oil price war. Therefore, even though Mexico is unlikely to join OPEC, as its share of the world's oil market grows to a level comparable to Iran's or Saudi Arabia's, it almost certainly will choose to adopt policies toward oil exports that increasingly complement the policies of OPEC's leaders. In sum, because expansion of Mexico's petroleum exports places increased constraints on OPEC's—especially Saudi Arabia's—economic, political, and strategic power, it will redound to the benefit of all the non-Communist, oil-importing countries. However, the United States will be deluding itself if it believes that the threat of greater Mexican petroleum exports can be used as an anti-OPEC weapon. Mexico has no animosity toward the OPEC countries (indeed, it has much to thank them for), and its policy makers appreciate the fact that important Mexican interests would be hurt by anti-OPEC policies. Therefore, the United States should be aware that any attempt to use Mexico as a tool for pursuing anti-OPEC ends might lead Mexico to respond by adopting a go-slow strategy toward the expansion of its petroleum industry. Since adoption of such a strategy would be to the detriment of both Mexico and the United States, every effort should be made to avoid this outcome.

NOTES

1. Edward Mitchell, *U.S. Energy Policy: A Primer* (Washington, D.C.: American Enterprise Institute, 1974), p. 5.

2. Bernardo F. Grossling, "Window on Oil Chart," testimony before the U.S., Congress, 95th Cong., 2d sess., Joint Economic Committee Subcommittee on Energy, March 1978; Bruce C. Netschert, "The Cloud on OPEC's Horizon," *Wall Street Journal*, March 29, 1976.

3. "World Oil Flow Slumps, OPEC Crude Crowded Out," *Oil and Gas Journal*, May 29, 1978, p. 21. Actual OPEC production was 31.5 million barrels per day; domestic consumption was less than 2 million barrels per day.

4. Organisation for Economic Cooperation and Development, *World Energy Outlook* (Paris: OECD, 1977), p. 86.

BIBLIOGRAPHY

BOOKS AND CHAPTERS IN BOOKS

Adelman, M. A. *The World Petroleum Market*. Baltimore: Johns Hopkins Press, 1972.

Adelman, M.A., ed. *Alaskan Oil: Cost and Supply*. New York: Praeger, 1971.

Allen, Loring. *Venezuelan Economic Development: A Political-Economic Analysis*. Greenwich, Conn.: JAI Press, 1977.

Association of Oil Producers in Mexico. *Documents Relating to the Attempt of the Government of Mexico to Confiscate Foreign-Owned Oil Properties*. February 1919.

Association of Producers of Petroleum in Mexico. *Documents with Regard to the Draft of the Expropriation Law*. Mexico City: APPM, 1936.

Baker, Roy Stannard. *Woodrow Wilson, Life and Letters*. Garden City, N.Y.: Doubleday, Doran, 1931.

Baruch, Bernard M. *Baruch, The Public Years*. New York: Holt, Rinehart & Winston, 1960.

Bell, Edward I. *The Political Shame of Mexico*. New York: McBride, Nast, 1914.

Bermúdez, Antonio J. *The Mexican National Petroleum Industry: A Case Study in Nationalization*. Palo Alto, Calif.: Institute of Hispanic American and Luso-Brazilian Studies, Stanford University, 1963.

Blustein, Howard I. *Venezuela, Area Handbook, 1977*. Washington, D.C.: Foreign Advisory Service of American University, 1977.

Bragaw, Louis. *The Challenge of Deepwater Terminals*. Lexington, Mass.: D. C. Heath, 1975.

Bullard, Fredda Jean. *Mexico's Natural Gas*. Austin: University of Texas Press, 1968.

Calvert, Peter. *The Mexican Revolution, 1910–1914*. Cambridge: Cambridge University Press, 1968.

Campos, Armand Maria Y. *Mugica: Cronica biografica*. Mexico City, CEPSA, 1939.

Churchill, Randolph Spencer, ed. *Companion Volume II, Part 3. Boston: Houghton-Mifflin, 1969.*

Churchill, Winston S. *The World Crisis, 1911–1914.* New York: Charles Scribner's Sons, 1930.

Cline, Howard F. *The United States and Mexico.* Cambridge, Mass.: Harvard University Press, 1953.

Cronon, David. *The Cabinet Diaries of Josephus Daniels, 1913–1921.* Lincoln: University of Nebraska Press, 1963.

Hendrick, Burton J. *Life and Letters of Walter H. Page.* Garden City, N.Y.: Doubleday, Page, 1925.

Herzog, Jesús Silva. *Petroleo Mexicano.* Mexico City: Fondo de Cultura Económica, 1941.

Hidy, Ralph W., and Hidy, Muriel E. *Pioneering in Big Business, 1882–1911.* New York: Harper & Brothers, 1955.

Howland, Charles P. *Survey of American Foreign Relations.* New Haven, Conn.: Yale University Press, 1931.

International Nominal Freight Scale Association. *Worldwide Tanker Nominal Freight Scale "Worldscale"* (as revised effective January 1, 1978).

Mancke, Richard B. *The Failure of U.S. Energy Policy.* New York: Columbia University Press, 1974.

Maull, Hans. *Oil and Influence: The Oil Weapon Examined.* Adelphi Paper, no. 117. London: International Institute for Strategic Studies, 1975.

Meyer, Lorenzo. *Mexico and the United States in the Oil Controversy.* Austin: University of Texas Press, 1972.

Mitchell, Edward. *U.S. Energy Policy: A Primer.* Washington, D.C.: American Enterprise Institute, 1974.

Organisation of Economic Cooperation and Development. *World Energy Outlook.* Paris: OECD, 1977.

Pan American Petroleum and Transport Company. *Mexican Petroleum.* New York: PAP & TC, 1922.

Pound, Arthur, and Moore, Samuel Taylor, eds. *The Told Barron: Conversations and Revelations of an American Pepys in Wall Street.* New York: Harper & Brothers, 1930.

Powell, J. Richard. *The Mexican Petroleum Industry: 1938–1950*. Berkeley: University of California Press, 1956.

Prewett, Virginia. *Reportage on Mexico*. New York: E. P. Dutton, 1941.

Purcell, John F. H., and Purcell, Susan Kaufman. "Matchine Politics and Socio-economic Change in Mexico." *In Contemporary Mexico*, edited by James W. Wilkie, Michael C. Meyer, and Edna Monzon de Wilkie, p. 366. Berkeley: University of California Press, 1976.

Purcell, Susan Kaufman. *The Mexican Profit-Sharing Decision*. Berkeley: University of California Press, 1975.

Rifai, Taki. *The Pricing of Crude Oil*. New York: Praeger, 1975.

Rippy, J. Fred; Vasconcelos, José; and Stevens, Guy. *American Policies Abroad*, University of Minnesota Press, 1959.

Rippy, J. Fred; Vasconcelos, José; and Stevens; Guy. *American Policies Abroad, Mexico*. Chicago: University of Chicago Press, 1928.

Sands, William Franklin. *Our Jungle Diplomacy*. Chapel Hill: University of North Carolina Press, 1944.

Schlagheck, James L. *The Political, Economic and Labor Climate in Mexico*. Philadelphia: Wharton School, University of Pennsylvania, 1977.

Skinner, Walter R. *Oil and Petroleum Manual, 1922*. London: 1922.

Spender, J. A. *Weetman Pearson, First Viscount Cowdray*. London: Cassell, 1930.

Standard Oil of New Jersey. *Confiscation or Expropriation? Mexico's Seizure of the Foreign-Owned Oil Industry*. New York: SONJ, 1940.

——— . Denials of Justice. New York: SONJ, 1938–40.

Tischendorf, Alfred. *Great Britain and Mexico in the Era of Porfirio Diaz*. Durham, N.C.: Duke University Press, 1961.

Tumulty, J. P. *Woodrow Wilson As I Know Him*. Garden City, N.Y.: Doubleday, Page, 1921.

von Sauer, Franz A. *The Alienated Loyal Opposition*. Albuquerque: University of New Mexico Press, 1974.

Wilkins, Mira. *The Emergence of Multinational Enterprise: American Business Abroad from the Colonial Era to 1914*. Cambridge, Mass.: Harvard University Press, 1970.

Winters, Toby. *Deepwater Ports in the U.S.* New York: Praeger, 1975.

GOVERNMENT DOCUMENTS

U.S., Department of Commerce. *Venezuela: A Survey of United States Business Opportunity.* Country Market Sectorial Survey, June 1976.

U.S., Congress, Senate. Committee on Foreign Relations. *Investigation of Mexican Affairs.* 66th Cong., 2d sess.

————. Subcommittee of the Committee on Foreign Relations. *Revolutions in Mexico.* 62 Cong., 2d sess., 1913.

U.S., Executive Office of the President. *The National Energy Plan.* Washington, D.C.: Government Printing Office, 1977.

U.S., President. Cabinet Task Force on Oil Import Controls. *The Oil Import Question.* Washington, D.C.: Government Printing Office, 1970.

MAGAZINE, NEWSPAPER, AND JOURNAL ARTICLES

"Aramco Sees Rapid Restoration of Crude." *Oil and Gas Journal*, January 28, 1974, p. 94.

"Brazil." *Seatrade*, January 1978, p. 81.

"Builders and Repairers Look to New Facilities." *Seatrade*, February 1978, p. 123.

"Burmah's Bahamas Terminal." *Seatrade*, January 1978, p. 24.

"Caribbean Report." *Seatrade*, February 1978, p. 139.

"Down Mexico Way." *Seatrade*, December 1977, p. 7.

Economist Intelligence Unit. *Quarterly Economic Review of Mexico*, 1977–78.

Fagen, Richard R. "The Realities of U.S./ Mexican Relations." *Foreign Affairs*, July 1977, p. 694.

"5,000 Persons Demonstrate Against Oil, Gas Sale to US." *Foreign Broadcast Information Service*, March 21, 1978, p. M1.

Franco, Alvaro. "Bay of Campeche May Rival Reforma Area." *Offshore*, January 1978, p. 43.

————. "Giant New Trend Balloons SE Mexico's Oil Potential." *Oil and Gas Journal*, September 19, 1977, p. 84.

———. "Latin America's Petroleum Surge Gathers Momentum." *Oil and Gas Journal*, June 5, 1978, p. 69.

———. "Mexico's Crude-Exporting Role May Be Short-Lived." *Oil and Gas Journal*, May 26, 1975, p. 27.

———. "Pemex Optimizes Reforma Operations." *Oil and Gas Journal*, March 28, 1977, pp. 135–36.

———. "Pemex Sees Reforma Extension Offshore." *Oil and Gas Journal*, March 7, 1977, p. 78.

"The Future for Mexico." *Euromoney* (supp.). April 1978.

"The Future for Money." *Euromoney* (supp.), April 1978.

Gordon, David. "Clouds of People." *Economist*, April 22, 1978, p. 7.

———. "Mexico, a Survey." *Economist*, April 22, 1978, p. 16.

Grayson, George W. "The Oil Boom." *Foreign Policy*, Winter 1977–78.

Huey, John. "Despite Rising Wealth in Oil, Mexico Battles Intractable Problems." *Wall Street Journal*, Ausust 30, 1978, p. 1.

———. Mexico's Economic Ills Could Topple Coalition If Workers, Poor Rebel." *Wall Street Journal*, August 8, 1977, p. 1.

"Industry Spending in U.S. to Hit Record $28.9 Billion." *Oil and Gas Journal*, February 20, 1978.

"International Briefs." *Oil and Gas Journal*, February 20, 1978, p. 90.

Kemp, Geoffrey. "Scarcity and Strategy." *Foreign Affairs*, January 1978, p. 401.

Martin, Douglas. "Mexicans Count on Oil to Help Repay Debts and Bolster Economy." *Wall Street Journal*, October 26, 1977, p. 1.

Masters, John A. "Deep Basin Gas Trap, West Canada." *Oil and Gas Journal*, September 18, 1978, pp. 226–41.

"Mexican Leader, Outlining Plans, Sees a Challenge in National Crisis." *New York Times*, February 1, 1978, p. 4.

"Mexico Expects U.S. to Buy Its Natural Gas Despite High Price." *Wall Street Journal*, March 21, 1978, p. 18.

"Mexico Eyes Starved U.S. as Outlet for Surplus." *Oil and Gas Journal*, June 27, 1977, p. 63.

"Mexico Grapples with Its Oil Bonanza." *New York Times*, May 7, 1978, p. F-3.

"Mexico Speeds Work on Huge Gas Line." *Oil and Gas Journal*, June 5, 1978, p. 121.

"Mexico: Tethered Watchdog." *Latin American Political Report*, October 14, 1977, p. 317.

"Mexico to Wait Rather Than Cut Gas Price to U.S." *New York Times*, January 6, 1978, p. D-5.

"Mexico's Oil, Foreign-Exchange Reserves Soar; President Cites Economic Recovery." *Wall Street Journal*, September 8, 1978, p. 6.

Meyerhoff, A. A., and Morris, A. E. L. "Central American Petroleum Potential Centered Mostly in Mexico." *Oil and Gas Journal*, October 17, 1977, pp. 104–09.

Nelson, Patricia. "Mexico Promoting Industry." *Journal of Commerce*, December 20, 1977, p. 9.

Netschert, Bruce C. "The Cloud on OPEC's Horizon." *Wall Street Journal*, March 29, 1976.

Nulty, Peter. "When We'll Start Running Out of Oil." *Fortune*, October 1977, pp. 247–48.

"Oil Gives Mexicans a Boost, But They Plan to Stay Out of OPEC." *Wall Street Journal*, February 11, 1978, p. 1.

"Pemex Director Talks About Oil Reserves." *Foreign Broadcast Information Service*, November 1, 1977, p. M2.

"Pemex has New Chiapas-Tabasco Finds." *Oil and Gas Journal*, May 2, 1977, p. 120.

"Pemex to Brown & Root: Yankees, Come In." *Forbes*, August 15, 1977, p. 28.

"President Defends Planned US Gas Pipeline." *Foreign Broadcast Information Service*, October 6, 1977, p. M1.

Riding, Alan. "Congress in Mexico Will Get New Power." *New York Times*, September 18, 1977, p. 20.

——. "Mexican Concerned that Reliance on Oil May Aggravate Ills." *New York Times*, December 31, 1977, p. 23.

——. "Silent Invasion: Why Mexico Is an American Problem." *Saturday Review*, July 8, 1978, p. 15.

Sanderman, Hugh. "Pemex Comes Out of Its Shell." *Fortune*, April 10, 1978, p. 18.

"Selected Crude Oil Prices." *Petroleum Economist*, June 1978, p. 273.

"Survey of Mexico." *The Economist* (supp.), April 22, 1978.

"Tipro Wants to Intervene in Mexican Gas Case." *Oil and Gas Journal*, September 12, 1977. p. 58.

Wilson, Howard M. "Plans Shaping Up for Alaska Gas Line." *Oil and Gas Journal*, April 3, 1978, p. 43.

UNPUBLISHED PAPERS

Fagen, Richard R. and Nau, Henry R. "Mexican Gas: The Northern Connection." Paper Read at Conference on the United States, U.S. Foreign Policy and Latin American and Caribbean Regimes, Washington, D.C., March 27–31, 1978.

ABOUT THE AUTHOR

RICHARD B. MANCKE is Associate Professor of International Economic Relations at the Fletcher School of Law and Diplomacy, Tufts University. His publications on energy policy issues include *The Failure of U.S. Energy Policy* (Columbia University Press, 1974), *The Performance of the Federal Energy Office* (American Enterprise Institute, 1975), and *Squeaking By: U.S. Energy Policy Since the Embargo* (Columbia University Press, 1976).

RELATED TITLES
Published by
Praeger Special Studies